D1580632

COLLINS GEM
BASIC FACTS

CHEMISTRY

W.A.H. Scott BSc PhD

Revised by
D.K. Jeffrey C.Chem. MRSC

COLLINS
London and Glasgow

First published 1982
Revised edition 1988
Reprint 10 9 8 7 6 5 4 3 2

© Wm. Collins Sons & Co. Ltd. 1988

ISBN 0 00 459102 10

Printed in Great Britain

Introduction

Basic Facts is a new generation of illustrated GEM dictionaries in important school subjects. They cover all the important ideas and topics in these subjects up to the level of first examinations.

Bold words in an entry means a word or idea developed further in a separate entry: *italic* words are highlighted for importance.

Elements are presented in this consistent format:

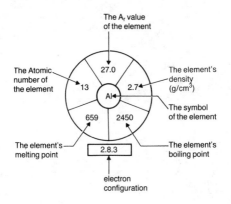

A_r The symbol for **relative atomic mass**. The A_r value for an **element** is a measure of the mass of an average atom of the element compared to the mass of an atom of the **isotope** carbon-12. See **M_r** and **Atomic Mass**.

Absolute temperature See **Kelvin temperature scale**.

Abundance The measure of how much of a substance exists. It is often expressed as a percentage. *Elements:* The abundances of **elements** in the earth and in the sun are shown in the tables below.

Earth elements	O	Si	Al	Fe	Ca	Na	K	Mg
% abundance	46.6	27.7	8.1	5.0	3.6	2.8	2.6	2.1

Sun elements	H	He	O	C	Si
% abundance	54.0	44.7	0.8	0.4	0.05

Isotopes: The abundances of **isotopes** in some commonly occurring elements are shown in the following table (continued over page).

element	% abundance of each isotope	
Chlorine	$^{35}_{17}\text{Cl}$ [75.5%]	$^{37}_{17}\text{Cl}$ [24.5%]
Carbon	$^{12}_{6}\text{C}$ [98.9%]	$^{13}_{6}\text{C}$ [1.1%]

1

| Magnesium | $^{24}_{12}$ Mg [78.6%] | $^{25}_{12}$ Mg [10.1%] |
| | | $^{26}_{12}$ Mg [11.3%] |

Knowing the percentage abundance of each isotope it is possible to calculate the A_r value for the element, e.g.:

$$^{35}_{17} \text{Cl} : ^{37}_{17} \text{Cl} = 3:1$$

$$A_r \text{ (Cl)} = \frac{(3 \times 35) + (1 \times 37)}{4} = 35.5$$

Accumulator This is a rechargeable **battery**. The most common type is the lead-acid accumulator. This is the battery which is used to start cars and lorries. It is also used to drive milk floats and fork-lift trucks. In the future we shall probably drive electric cars which are powered by accumulators.

In the lead-acid accumulator the **electrolyte** is dilute **sulphuric acid**. The positive plate is made of lead(IV) oxide and the negative plate is composed of lead. When the accumulator is discharged (flat) the plates are coated with lead(II) sulphate. The sulphate coating is removed when the accumulator is recharged by passing electricity through it. The lead-acid accumulator is dangerous because it can produce very high currents. It produces the flammable gas **hydrogen** whilst it is being recharged and there is the constant danger of the **acid** spilling from the accumulator.

2

The nickel-cadmium accumulator is 'dry'. The positive plate is made of a complex nickel salt (nickel oxyhydroxide), the negative plate is made up of cadmium. The electrolyte is **potassium hydroxide** solution which is soaked into a spongy material and so cannot spill. Nickel-cadmium accumulators work well at low temperatures but, as yet, are very expensive.

Acid A substance which releases **hydrogen ions** (H^+) when it is added to **water**. The hydrogen ion is solvated, i.e., a water **molecule** adds on to it, to give the **oxonium ion** (H_3O^+). Acidic solutions have a **pH** of less than 7.

Common laboratory acids are:

> **Nitric acid** HNO_3
> **Hydrochloric acid** HCl
> **Sulphuric acid** H_2SO_4
> **Ethanoic acid** CH_3COOH

These acids are dangerous, corrosive liquids and should always be treated with care.

Acids:
★ turn *blue* **litmus** *red*
★ give **carbon dioxide** when added to **carbonates**
★ give **hydrogen** when added to certain **metals**
★ neutralize **alkalis**

Acid-base reactions Acids react with *bases* to form a *salt* and water only. Some examples are shown here.

Hydrochloric acid HCl(aq)	+	Sodium hydroxide NaOH(aq)	→	Sodium chloride NaCl(aq)	+	water H_2O(l)

3

Nitric acid	+ Copper(II)	→ Copper(II)	+ water
$2HNO_3(aq)$	oxide	nitrate	$H_2O(l)$
	$CuO(s)$	$Cu(NO_3)_2(aq)$	

Acid salts Salts in which only some of the replaceable **hydrogen atoms** in an **acid molecule** have been replaced by a metal. Some examples are shown below:

Na_2CO_3	Sodium carbonate	$NaHCO_3$	Sodium hydrogencarbonate
Na_2SO_4	Sodium sulphate	$NaHSO_4$	Sodium hydrogensulphate
Na_3PO_4	Sodium phosphate	Na_2HPO_4	Sodium hydrogenphosphate

Acidic oxides Most **oxides** of **nonmetals** react with water to form acidic **solutions.** Because of this they are known as acidic oxides. Examples:

oxide		acid	
Carbon dioxide	CO_2	Carbonic acid	H_2CO_3
Sulphur dioxide	SO_2	Sulphurous acid	H_2SO_3
Sulphur trioxide	SO_3	Sulphuric acid	H_2SO_4

Activation energy (E_a) When **hydrogen** and **oxygen** are mixed there is no **reaction**. When a **flame** is brought into contact with the gases, there is an immediate **explosion**. The heat of the flame causes the reaction to occur. Many chemical reactions are like this. It seems that there is a barrier which has to be overcome before the reaction will take place. **Energy**

4

(e.g. the heat of the flame) has to be put in to make the reaction occur. This energy is known as the activation energy. See **Catalyst**.

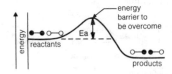

Addition reaction A reaction in which two or more molecules are reacted together to form a single molecule. Good examples are found in the reactions of **alkenes**. These compounds contain **double bonds** and can add an atom to each side of the double bond to form a **saturated alkane**. The example of **ethene** reacting with **hydrogen** is shown here.

ethene + hydrogen⟶ ethane

Addition reactions also occur in the making of many **polymers**, e.g. **poly(ethene)**.

Air The **mixture** of gases which surrounds the earth. The average composition of pure air is shown below. This composition varies from place to place and also varies with altitude.

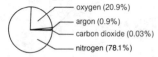

oxygen (20.9%)
argon (0.9%)
carbon dioxide (0.03%)
nitrogen (78.1%)

Other components
neon, helium, krypton, xenon — In small constant amounts.
water — In very variable amounts.

Air also contains pollutants, some of which are shown below.

pollutant	sources
Sulphur dioxide	Burning coal, oil.
Carbon monoxide	Engines, cigarettes.
Oxides of nitrogen	Car engines.
Soot	Fires, engines.
Pollen	Trees, flowers.
Dust	Volcanoes.

Air is vital for life. The **oxygen** is necessary for **respiration** and the **carbon dioxide** needed for **photosynthesis**.

Alcohols Alcohols are important **organic** compounds. They possess the arrangement of atoms:

$$-\overset{|}{\underset{|}{C}} - O - H$$

The —OH group is the **functional group** of the alcohols.

Ethanol is the most important compound but see also **methanol** and **glycol**.

Alcohols with a small M_r value are flammable liquids which dissolve in water. Some of the important properties of alcohols are that they can be oxidized easily, they form **esters** and react with **sodium** to produce **hydrogen**.

Alkali   A base which is soluble in water. They are usually metal **hydroxides**, e.g. **sodium hydroxide**, but **ammonia** solution is also an alkali. Their reactions are affected by their *strength* and their **concentration**.

Oven cleaners, household **ammonia** and some paint strippers contain alkalis.

Alkalis:
★ turn red **litmus** blue
★ neutralize **acids**
★ have a **pH** above 7
★ react with acids to produce a **salt** and **water** only.

Alkanes   Alkanes are **hydrocarbons**. They are very important compounds in our lives. Our gas supplies (North Sea gas and 'Calor' gas) are almost 100% alkanes. Vaseline or petroleum jelly is made of alkanes. The chief source of alkanes is **petroleum** and they are valuable raw materials in the chemical industry.

They have a **general formula:** CH_nH_{2n+2} so the

alkane with 4 carbon atoms will have 10 hydrogen atoms: C_4H_{10}, **butane**.

See also **Methane, Ethane** and **Propane**. They are **saturated** compounds and so are not very reactive. They tend to be **flammable** and will react with **chlorine** in **ultraviolet radiation**.

Alkenes Alkenes are important **hydrocarbon** compounds. They are widely made in oil refineries and are used as starting materials in the manufacture of many materials, e.g. **plastics**. **Ethene**, propene and styrene (phenylethene) are three of the most important alkenes.

They have a **general formula**: C_nH_{2n}. So the alkene with 3 carbon atoms (propene) will have 6 hydrogen atoms. Because they are **unsaturated** compounds they have a carbon-carbon **double bond** and are reactive compounds. Typically they will undergo **addition reactions** with hydrogen, **halogens** and water. They will also form **polymers** by this addition process.

See **Poly(phenylethene)** and **Poly(propene)**.

Alkynes There is only one common alkyne. This is **ethyne**. Alkynes have the **general formula** C_nH_{2n-2} and so, ethyne with 2 carbon atoms will have 2 hydrogen atoms: C_2H_2. Alkynes have carbon-carbon **triple bonds** and therefore show the typical **addition reactions** of **unsaturated** compounds.

Allotropes Some elements exist in different forms

8

in the same physical state. The chemical properties are the same but the physical **properties** are different. The best example is carbon. There are two allotropes: **diamond** and **graphite**. Diamond is hard and colourless, graphite is flakey and black. Other elements which exist in different allotropic forms are **sulphur**, **phosphorus** and **tin**.

Alloy An alloy is a **mixture** which is made up of two or more **metals** or which contains metals and **nonmetals**.

Alloys are much more widely used than pure metals because, by bringing two or more elements together in the right proportions, specific properties can be obtained. For example, aluminium is quite a soft metal but when a small amount of copper is mixed in, the alloy duralumin is produced which is strong enough to be used in aircraft frames. Some common examples are: **brass, bronze, duralumin, pewter, solder** and **steel**.

Alpha particles ($^4_2He^{2+}$) These are **helium** atoms without their electrons. They are produced in many **nuclear reactions**. They have a fairly short range in air (usually less than 10 cm) and are easily stopped by thin sheets of paper or foil.

Aluminium Aluminium is the most abundant **metal** in the earth's crust (approximately 8 per cent by mass). Its **alloys** are widely used at home and in industry.

Clay, shale, slate and granite all contain aluminium compounds but the metal is difficult to extract from them. It is obtained by the **electrolysis** of **bauxite** dissolved in **cryolite**, using graphite electrodes.

The metal has **valency** 3. Although it is a very reactive metal, it, and its alloys, are resistant to **corrosion** because of a layer of **oxide** on the surface of the metal. It is subject to **anodizing**.

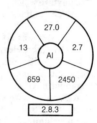

Aluminium compounds

Aluminium chloride $AlCl_3$	The **anhydrous** chloride is **covalent**.
Aluminium oxide Al_2O_3	The oxide is **amphoteric**.
Aluminium sulphate $Al_2(SO_4)_3$	This is the most important aluminium compound. It is used as a precipitator in sewage works, as a mordant and as a size in the paper industry. It is also used as a foaming agent in fire-extinguishers.

Amalgam An amalgam is an **alloy** which contains **mercury**. Zinc amalgam is used for teeth fillings. *Amalgamation* was once used to extract gold and silver from crushed rock or earth which contained the metals.

Amino acids These are **organic** compounds which contain —COOH and —NH$_2$ groups. They are the building bricks from which **proteins** are made. Twenty different amino acids are needed to make up all the proteins in our bodies. Some of these amino acids our bodies can produce but there are eight which we need to eat — the *essential* amino acids. Amino acids combine together to form **peptides**.

glycine

Ammonia (NH₃) A colourless gas with a pungent odour. It is very soluble in water giving an *alkaline* solution. Ammonia was once made from coal but now over 90% comes from the **Haber process**.

Ammonia is also produced by bacteria found on the roots of leguminous plants (peas and beans). When protein decomposes, ammonia is released. Both of these are important sources of plant food. See the **Nitrogen cycle**.

The importance of ammonia in our lives is shown in the diagram below.

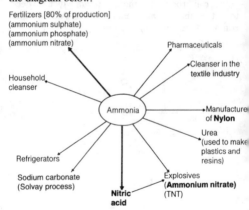

Fertilizers [80% of production]
(ammonium sulphate)
(ammonium phosphate)
(ammonium nitrate)

Pharmaceuticals

Cleanser in the textile industry

Household cleanser

Ammonia

Manufacture of **Nylon**

Urea
(used to make plastics and resins)

Refrigerators

Sodium carbonate
(Solvay process)

Nitric acid

Explosives
(**Ammonium nitrate**)
(TNT)

THE USES OF AMMONIA

Ammonia is a **covalent compound**. It has a characteristic reaction with **hydrogen chloride** to give dense white fumes of ammonium chloride:

$$NH_3(g) + HCl(g) \rightarrow NH_4Cl(s)$$

Ammonia can be oxidized to **nitric acid** and this is the major source of nitric acid today. It is easily liquefied and it is usually carried in this state from place to place in tankers.

Ammonia solution (NH$_4$OH) (formerly called **Ammonium hydroxide**) A solution of **ammonia** in water.

Ammonium compounds

Ammonium carbonate $(NH_4)_2CO_3$	An unstable compound used as smelling salts *sal volatile*.
Ammonium chloride NH_4Cl	A salt which undergoes **dissociation** into ammonia and hydrogen chloride.
Ammonium nitrate NH_4NO_3	A very important compound which is used as a fertilizer and an explosive.
Ammonium phosphate $(NH_4)_3PO_4$	A convenient chemical with which to put both nitrogen and phosphorus into the soil.
Ammonium sulphate $(NH_4)_2SO_4$	An important fertilizer.

Ammonium ion Ammonium compounds contain the ammonium ion as does a solution of ammonia gas in water. The ion is positively charged and the hydrogen atoms are arranged around the nitrogen atom in a **tetrahedral** way.

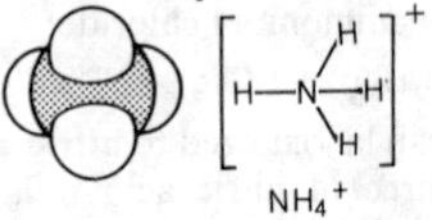

NH_4^+

Amphoteric (of oxides and hydroxides) Insoluble **oxides** and **hydroxides** which show both basic and acidic **properties** are said to be 'amphoteric'. The examples shown here will react with both **acids** and **alkalis** to form **salts**.

Examples		
Al_2O_3	PbO	ZnO
$Al(OH)_3$	$Pb(OH)_2$	$Zn(OH)_2$

For example:

$$ZnO + 2HCl \rightarrow ZnCl_2 + 2H_2O$$
$$ZnO + 2NaOH + H_2O \rightarrow Na_2Zn(OH)_4$$
$$\text{(sodium zincate)}$$

The solid **hydroxides** of these metals all dissolve in **sodium hydroxide** solution.

Amorphous Material which is non-crystalline, i.e. appears to have no regularity of shape is described as amorphous; a-morphous = without definite shape or form.

Ampère (amp) This is the **unit** of electric current. It measures the rate of flow of charge. One amp = 1 **coulomb/second**.

14

Anhydrides These are substances which react with water to produce acids. All **acidic oxides** are anhydrides, but some will give rise to one acid only whilst others will produce two.

Examples	
Anhydride	Acid
SO_3	H_2SO_4
NO_2	$\begin{cases} HNO_3 \\ HNO_2 \end{cases}$

Anhydrous Containing no water. The term usually denotes salts with no **water of crystallization**.

For example:

Anhydrous copper(II) sulphate $CuSO_4$
Anhydrous sodium carbonate Na_2CO_3

The **hydrated** salts are:

$CuSO_4 . 5H_2O$
$Na_2Cl_3 . 10H_2O$

The term is also used about liquids which are perfectly dry, e.g. anhydrous ether.

Anion A negatively charged **ion**. They usually contain a full outer orbital of electrons. Nonmetals form anions but complex ions can also be anions.

For example:

Sulphate SO_4^{2-} Manganate(VII) MnO_4^-
Carbonate CO_3^{2-} Nitrate NO_3^-

Annealing Heating and cooling **metals** causes

changes in their properties, especially their strength. When a metal is heated to a high temperature and then cooled *very* slowly the metal becomes soft and **malleable**. This is because the **crystals** in the metal have been allowed to grow very large. This slow-cooling is called 'annealing'.

Anode When a solution undergoes **electrolysis** the **electrode** with the *positive* potential is called the anode. **Anions** are attracted to the anode because they are negatively charged. The ions give up their extra **electrons** at the anode. Nonmetallic elements are produced and the electrons travel round the circuit. See **Cathode**.

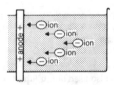

Anodizing Aluminium has a protective layer of oxide (Al_2O_3). This layer of oxide can be thickened by making the piece of aluminium the **anode** (+) in an **electrolysis** cell. Dilute **sulphuric** acid is used as the **electrolyte**. Aluminium atoms give up electrons and they react with the water:

$$2Al(s) + 3H_2O(l) \rightarrow$$
$$Al_2O_3(s) + 6H^+(aq) + 6 \text{ electrons}^-.$$

Aluminium oxide is formed and this can be dyed to produce attractive finishes to products, e.g. coloured milk-bottle tops.

Antifreeze This is a substance which is added to the cooling system of engines in winter to prevent the formation of **ice** which would damage the engine. The substances added *lower* the freezing point of the water. The two most commonly used substances are:

Methanol $CH_3 - OH$

Ethane $- 1, 2 -$ diol

$$HO \quad\quad OH$$
$$CH_2 - CH_2$$

aq This abbreviation is used to denote an **aqueous solution**, e.g. $NaCl(aq)$.

Aqueous solution This is a solution where **water** is the **solvent**.

Argon (Ar) The most abundant **noble gas**. It makes up 0.9% of the **atmosphere** (by volume). It is very unreactive. No **compound** of argon has yet been made.

Argon has uses where the presence of an unreactive gas is useful, e.g. light bulbs and fluorescent tubes. See diagram over page.

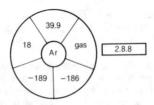

Aromatic (of compounds) Containing a **benzene** ring structure in their molecules.

Arsenic A brittle grey **metalloid**. Its compounds are very poisonous – especially the oxide (As_2O_3). It is a group v element which occurs as **allotropes**. Arsenic compounds are used in pesticides. The grey allotrope **sublimes** at 613°C.

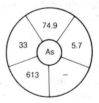

Asbestos A naturally occurring silicate which was widely used in the past for fireproof clothing and insulating materials. Its use today is restricted because of the dangers of contracting *asbestosis*, a fatal lung disease. It was much used in schools as bench mats and centre pieces for gauzes. These are now made of material which does not contain asbestos.

Atmosphere **1.** A unit of **pressure**. Although the pressure of the air which surrounds us varies from place to place and from time to time its value usually only varies slightly and is always approximately one atmosphere.

2. The mixture of gases which surrounds a planet, sun or moon. The atmosphere on earth is **air**.

Atom The smallest indivisible particle of an element that can exist, the building bricks with which everything is made. Atoms are small particles; 100 million placed end-to-end would measure 1 cm. They are made up of even smaller **subatomic particles**:

Protons (p)
Neutrons (n) These particles are found in the centre of the atom — the **nucleus**.

Electrons (e) These particles move round the nucleus.

The atom as a whole is electrically **neutral** although the protons and electrons carry electrical charges. These charges are equal in size but opposite in sign, hence the number of protons always equals the number of electrons. Atoms which lose or gain electrons are called **ions**. All atoms of the same **element** contain the same number of protons and have the same **atomic number**, but atoms of the same element can contain different numbers of neutrons. See **Isotope**.

An atom can be thought of as the smallest part of an element that can take part in a chemical reaction.

Atomic mass Atoms have different masses if they contain different numbers of **protons, neutrons** and **electrons**. Because these particles are so small the mass of an atom is tiny. It is not usual to refer to the mass of an atom in grams or kilograms so we normally compare the mass of one atom with a standard mass. This is called the relative atomic mass (A_r).

Atomic number (Z) The number of **protons** in the **nucleus** of an atom. All atoms of the same **element** have the same atomic number, e.g. sodium atoms all contain 11 protons. In a **neutral** atom the number of electrons equals the atomic number.

Atomicity The number of **atoms** in a **molecule** of an element. With the **inert gases** there is only one atom in the molecule and the atomicity is 1. All other gases have an atomicity of 2, e.g. **oxygen** (O_2), **nitrogen** (N_2), except **ozone** (O_3). Molecules in solid **sulphur** have an atomicity of 8 (S_8).

Avogadro constant/number (L) The number of

atoms that are contained in exactly 12 g of the carbon-12 **isotope**. More generally it is the number of particles present in a **mole** of substance. L = 6×10^{23} mol^{-1}.

Avogadro's hypothesis/law This stated that under the same conditions of temperature and pressure, equal **volumes** of gases contain the same number of **molecules**, e.g. 10 dm^3 of oxygen and 10 dm^3 of hydrogen both contain the same number of molecules. They contain twice as many molecules as 5 dm^3 of **chlorine** − provided that all the volumes were measured at the same temperature and pressure.

Baking powder A mixture of a carbonate and a weak acid which is used in cooking. When water is added to the solid mixture or the mixture is heated, **carbon dioxide** is produced. This produces bubbles in the dough or cake mixture and it *rises*, producing a product with an open texture. The mixture usually contains sodium hydrogencarbonate ($NaHCO_3$) and tartaric acid.

The powder can be made up in the kitchen by mixing cream of tartar (the potassium salt of tartaric acid) and baking soda (sodium hydrogencarbonate). This prevents the reaction from occurring before it is required.

Balance 1. A device for comparing the **masses** of objects. It is usual to use electric balances in schools these days.

2. Chemical **equations** must be 'balanced'. It is

21

necessary to have equal numbers of each atom on each side of the equation as atoms can neither be created nor destroyed.

For example:

$$Zn(s) + HCl(aq) \rightarrow ZnCl_2(aq) + H_2(g)$$

No. of atoms:

$$Zn = 1 \; H = 1 \; Cl = 1 \quad Zn = 1 \; Cl = 2 \; H = 2$$

The equation *is not* balanced.

$$Zn(s) + 2HCl(aq) \rightarrow ZnCl_2(aq) + H_2(g)$$

The equation *is* balanced.

Barium Barium is a very reactive **group II element**. It occurs in nature as the **sulphate** (barytes) and as the **carbonate**. The **metal** is extracted from the molten **chloride** by electrolysis.

Barium chloride solution is used to test for sulphates.

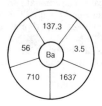

Barrel 1. A cask for holding beer. It has a **volume** of exactly 32 imperial gallons.

2. A unit volume in the **petroleum** industry. One

barrel = approximately 35 imperial gallons. (159 litres).

Base   A base is a substance which reacts with an acid to form a **salt** and **water** only. Bases are usually **metal oxides** or **hydroxides**, e.g. sodium hydroxide (NaOH) and copper (II) oxide (CuO).

$$NaOH(aq) + HCl(aq) \rightarrow NaCl(aq) + H_2O(l)$$
$$CuO(s) + H_2SO_4(aq) \rightarrow CuSO_4(aq) + H_2O(l)$$

Metal oxides and hydroxides which are soluble in water are known as **alkalis**. These are compounds of **group I** and **II** metals.

Ammonia solution is an alkali. It contains hydroxide ions in *equilibrium:*

$$NH_3(g) + H_2O(e) \rightleftharpoons \overset{+}{N}H_4(aq) + \overset{-}{O}H(aq)$$

Basic oxide   An oxide which reacts with an acid to form a **salt** and **water** only. Basic oxides are oxides of metals, but not all metals give basic oxides. See **Amphoteric oxide** and **Acidic oxide**.

Basic salts   Salts which contain **hydroxide ions** as well as normal **anions**, e.g. **sulphate**, **carbonate**.

Common examples are:

Malachite	$CuCO_3. Cu(OH)_2$
Azurite	$2CuCO_3. Cu(OH)_2$
White lead	$2PbCO_3. Pb(OH)_2$

Battery   A device which converts chemical **energy** into electrical energy. **Chemical reactions** occur and, in them, **electrons** are sent through a circuit. Batteries are of two types: the rechargeable and the non-rechargeable. The former kind are known as **accumulators**.

Batteries are available in different shapes, sizes and price ranges and their chemical composition varies too. The common 'dry battery' (the Leclanché dry cell) has **carbon** and **zinc electrodes**, and a gel of ammonium chloride as the **electrolyte**. It produces a voltage of 1.5 volts. See **Cell**.

Bauxite The chief **ore** from which **aluminium** is extracted. It is a **hydrated** oxide ($Al_2O_3 \cdot x H_2O$) and is found in tropical regions of the world, e.g. Northern Australia and West Africa.

Beehive shelf This piece of apparatus is used to allow a gas to be collected over water.

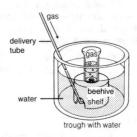

gas

delivery
tube

gas

beehive
shelf

water

trough with water

Benzene Benzene (C_6H_6) is the simplest **aromatic**

compound. It is a toxic liquid **hydrocarbon** which gives rise to the production of cancers. Its widespread use in schools has been replaced by that of **methylbenzene**.

Beta particles These are **electrons** which are produced in the following **nuclear reaction**:

$$_{0}^{1}n \rightarrow {}_{1}^{1}p + {}_{-1}^{0}e$$

neutron   proton   electron

E.g. $_{15}^{31}p + {}_{0}^{1}n \rightarrow {}_{15}^{32}p \rightarrow {}_{16}^{32}s + {}_{-1}^{0}e$

In this radioactive decay the electrons are expelled from the **nucleus** of the atom. They travel at high speeds (up to 98% of the speed of light) and have a greater penetrating power than **alpha particles**. When this reaction occurs, the atom concerned changes its **atomic number** because it gains a **proton**.

Bitumen A mixture of high boiling-point **hydrocarbons** which is left behind in the distillation of **petroleum**. It is used for roofing and road surfacing.

Blast furnace A furnace allowing a continuous production of molten iron. At the bottom of the furnace the coke and hot air react:

$$C(s) + O_2(g) \rightarrow CO_2(g)$$

This **exothermic** reaction raises the temperature to 1800°C. The coke then reacts with the carbon dioxide:

$$C(s) + CO_2(g) \rightarrow 2CO(g)$$

The carbon monoxide formed reduces the iron oxides to iron (1200°C). For example:

$$Fe_2O_3(s) + 3CO(g) \rightarrow 3CO_2(g) + 2Fe(l)$$

The molten iron flows to the bottom of the furnace. The **limestone** in the charge is decomposed by the heat producing calcium oxide and carbon dioxide.

The calcium oxide reacts with impurities from the iron ore (mainly silica (SiO_2)) and forms a molten slag:

$$CaO(s) + SiO_2(s) \rightarrow CaSiO_3(l)$$

This sinks to the bottom of the furnace where it floats on top of the iron. The iron which is produced contains about 3% **carbon** and the **metal** is brittle. In **steel**-making the amount of impurities in the iron is very carefully controlled. See diagram at top of next page.

Bleach A substance used to decolourize materials, e.g. fabrics and paper. Sunlight and oxygen act as bleaches but the most commonly found bleach is sodium chlorate(I) solution (NaClO). This is produced

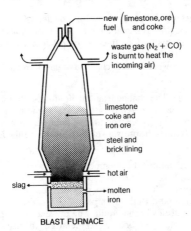

new fuel (limestone, ore and coke)

waste gas (N_2 + CO) is burnt to heat the incoming air)

limestone coke and iron ore

steel and brick lining

hot air

slag

molten iron

BLAST FURNACE

when **chlorine** is reacted with **sodium hydroxide** solution:

$$Cl_2(g) + 2NaOH(aq) \rightarrow$$
$$NaCl(aq) + NaClO(aq) + H_2O(l)$$

Domestic bleaches such as 'Domestos' contain sodium chlorate(I) solution. The chlorate(I) decomposes to give oxygen which acts as an oxidizing agent:

$$
\begin{array}{ccc}
NaClO + coloured & \rightarrow & NaCl + oxidized \\
material & & (decolourized) \\
& & material
\end{array}
$$

Bleaching powder A white powder of complex composition. It is made by the action of **chlorine** on calcium hydroxide. It contains calcium chlorate(I) $Ca(OCl)_2$.

The powder releases chlorine when it is treated with dilute acid.

Boiling The change from the **liquid** state to the **gas** state at a fixed **temperature** — the **boiling point**. Boiling takes place when the vapour pressure of the liquid equals the pressure of the atmosphere above the liquid.

Boiling point The temperature at which **boiling** occurs. It is not fixed. It depends on the atmospheric **pressure**. The higher the pressure, the higher the boiling point. Boiling temperatures are also affected by impurities in the liquid. The presence of impurities causes the boiling temperature to rise. See **Evaporation**.

Bond **Atoms** are held together in **molecules** and **giant structures** by chemical bonds. These bonds are the forces which exist between the atoms and are generated by **electrons**. Electrons are either *shared between* atoms, or atoms *gain* or *lose* electrons forming ions.

The sharing of electrons (**covalent bonding**) usually occurs when two nonmetallic atoms join together, e.g. H—H or H—Cl. The loss or gain of electrons (**ionic** or electrovalent bonding) occurs when atoms of metallic

elements join with those of nonmetallic elements, e.g.
Na + Cl or Mg + O.

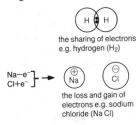

the sharing of electrons
e.g. hydrogen (H$_2$)

Na$-$e$^-$
Cl$+$e$^-$ $\}$ $\rightarrow$

the loss and gain of
electrons e.g. sodium
chloride (Na Cl)

Boron A nonmetallic element (see **Nonmetal**) in group III of the **periodic table**. Its oxide is used to make glass — see **Borosilicate glass**. It is a component of **alloys** and is a moderator in nuclear reactors.

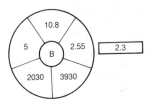

Borosilicate glass The addition of boron oxide (B$_2$O$_3$) during the manufacture of **glass** results in a material which does not expand very much. Con-

29

sequently the glass can be heated or cooled rapidly without cracking. 'Pyrex' glassware is like this.

Boyle's law At constant temperature the **volume** of a fixed **mass** of gas is inversely proportional to the **pressure** of the gas, i.e. if the pressure is *doubled* the volume is *halved*.

$$P \propto \frac{1}{V}$$

$$P.V. = \text{constant}$$

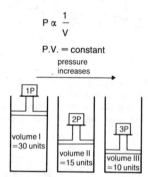

Brass Brass is an **alloy** of **copper** and **zinc** which usually contains about 30% zinc. Brass is an attractive yellow-golden coloured material which is used for ornaments and for electrical components, e.g. in electrical plugs.

Brine A solution of sodium chloride in water. The solution is more concentrated than seawater and is used in the food processing industry, e.g. the pro-

30

duction of bacon and in the preservation of many foods. In the chemical industry **chlorine** is produced by the **electrolysis** of brine.

Bromide A **compound** of **bromine** and another **element**.

Metal bromides are usually **ionic** solids, e.g. sodium bromide NaBr. See **Halides**.

Nonmetal bromides are usually **covalent** compounds, e.g. hydrogen bromide HBr.

Bromine (Br_2) A member of the **halogen** group of elements. It is a **volatile** red liquid at room temperature. The liquid is very corrosive and the vapour is both irritating and poisonous.

Although it is less reactive than **chlorine**, bromine will react vigorously with metals forming **bromides**:

$$Mg(s) + Br_2(l) \rightarrow Mg\ Br_2(s)$$

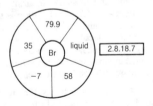

Bromine is extracted from seawater by treatment with **chlorine**. One thousand litres of seawater contains about 75 g of bromine. Its chief uses are in the production of silver bromide for use in photographic films and in the manufacture of 1, 2– dibromoethane which is added to **petrol** to prevent engine cylinders being blocked by deposits of lead. See **Octane rating**.

1,2 dibromoethene

Bronze The combination of **copper** ($>90\%$) with **tin** ($<10\%$) results in an **alloy** which is much stronger than copper. The discovery of bronze in the Middle East (3000 B.C.) gave rise to important changes in the way man lived (The Bronze Age). Nowadays, bronze is mainly used for gear wheels and engine bearings and, of course, as medals e.g. for the Olympic Games.

Brownian motion Particles suspended in a **liquid** or **gas**, e.g. pollen in water or smoke in air, are seen to move in a random way. The explanation is that the microscopic particles which make up the liquid or gas, e.g. water and air **molecules**, are moving randomly and are hitting against the larger particles of pollen or smoke, causing them to move. Brownian motion is put forward as evidence for the **kinetic theory**.

Brown ring test A test for **nitrates**. If **concentrated sulphuric acid** is carefully poured into an **aqueous solution** of a nitrate to which an acidic solution of iron(II) sulphate has been added, a brown ring forms at the junction of the two layers of liquid. The nitric acid which is formed reacts with the iron(II) sulphate and produces a brown compound.

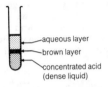

aqueous layer
brown layer
concentrated acid (dense liquid)

Burette A long glass tube with a tap on the end. The tube is marked (graduated) every 1 cm³ with smaller graduations of 0.1 cm³. It is used in **titrations** to measure accurately small volumes (up to 50 cm³) of liquid. The volume under the last graduation is called the dead space because it is not used for measurements. See **Pipette**.

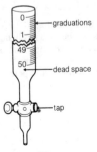

graduations

dead space

tap

Burner (or **Bunsen burner**) A device invented in the nineteenth century so that coal-gas could be burnt cleanly giving a hot flame. **Natural gas** is now used. With the air hole open the gas is mixed with air and this makes sure that all the gas is burnt and no soot is produced. With the air hole shut a yellow sooty flame is produced because the gas is not fully burnt. See **Flame**.

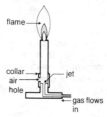

Burning See **Combustion**.

Butane Butane is an **alkane** compound. It is a gas at room temperature but it is easily liquefied. It is present (dissolved) in **petroleum** and is used in refineries both as a fuel and as a starting material for the production of **hydrogen** and **petrol** components.

It is a major component of 'Calor gas'. The gas burns in air:

$$2C_4H_{10}(g) + 13O_2(g) \rightarrow 8CO_2(g) + 10H_2O(g)$$

See diagram at top of next page.

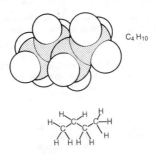

$C_4 H_{10}$

By-product A by-product of a reaction is something which is produced in addition to the product which is required. For example, slag is a by-product of making **iron** in the **blast furnace** and **carbon dioxide** is a by-product of the **fermentation** process. It is important, if processes are to be economic, that by-products are sold, e.g. slag is sold as road making material.

Calcium A soft, metallic **element** which is in **group** II

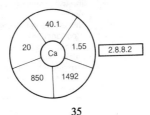

40.1

20

Ca

1.55

2.8.8.2

850

1492

of the **periodic table**. It is a fairly reactive **metal**, giving a slow but steady stream of **hydrogen** when it is added to cold water. 3.6% of the earth is made up of calcium compounds. Many of these are very common, e.g. **limestone** and **chalk** ($CaCO_3$). Calcium is obtained by the **electrolysis** of molten calcium chloride.

Calcium compounds

Calcium carbonate $CaCO_3$	This is used as building stone and in the production of **cement. Lime** (calcium oxide) is produced from calcium carbonate in a *lime kiln*. See **Hardness of water**.
Calcium chloride $CaCl_2$	Used to produce calcium metal. The **anhydrous salt** is used as a drying agent.
Calcium hydrogen-carbonate $Ca(HCO_3)_2$	This is an important cause of temporary **hardness** in water.
Calcium hydroxide (slaked lime) $Ca(OH)_2$	This is a slightly soluble alkali. The aqueous solution is called **limewater**. It is used to make mortar.
Calcium oxide CaO	This is **lime**. It has very important uses in agriculture to combat excess acidity in soil.
Calcium sulphate $CaSO_4$	Calcium sulphate gives rise to permanent hardness in water. It is found as anhydrite and gypsum.

Calor gas A mixture of **hydrocarbon** compounds which is used as a portable supply of gas. It is made up of mainly **propane** (C_3H_8) and **butane** (C_4H_{10}) and these compounds are stored in metallic bottles under **pressure**.

Carbohydrate **Organic** compounds which contain the **elements, carbon, hydrogen** and **oxygen** only. Their formulae are always of the form: $C_x(H_2O)_y$, e.g. **sugars** such as **glucose** ($C_6H_{12}O_6$) **sucrose** ($C_{12}H_{22}O_{11}$) and ribose ($C_5H_{10}O_5$) and **polymers** such as **starch** ($C_6H_{10}O_5$)$_n$ and **cellulose** ($C_6H_{10}O_5$)$_n$. See **Monosaccharide, Disaccharide** and **Polysaccharide**.

Carbon A nonmetallic element (see **Nonmetal**) which is in **group** (IV) of the **periodic table**. It is found in nature as two **allotropes: diamond** and **graphite**.

All living tissue contains carbon compounds (**organic** compounds), e.g. **carbohydrates, proteins,** fats, and without these compounds life would be impossible.

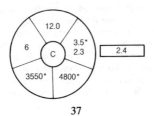

Although the element burns in **oxygen** or **air** it is otherwise fairly unreactive. It is an important reducting agent. **Coke, coal** and **charcoal** are all impure forms of carbon. Carbon is also found combined in materials such as **limestone**, in the form of metal carbonates.

Carbon compounds

Carbon dioxide CO_2
    This gas can be formed by the action of heat or **acids** on **carbonates** or by the complete combustion of carbon. It is also produced in the **fermentation** of **sugars**. The test for carbon dioxide is that it turns **limewater** milky. Carbon dioxide plays a vital role in **photosynthesis** and the **carbon cycle**.
Carbon monoxide CO
    This gas is formed when carbon or its compounds are not completely burnt. It is an **air** pollutant, being produced in internal combustion engines and from the burning of cigarettes. It is a very poisonous gas because it combines with the **haemoglobin** in blood. It is useful as a **reducing agent** in the **blast furnace**. The gas burns to give **carbon dioxide**.

Carbon cycle Every time an animal breathes or a fuel is burnt carbon dioxide is released into the **atmosphere**. The process of **photosynthesis** continually takes carbon dioxide out of the atmosphere and builds up plant structures with it. In these processes carbon dioxide is cycled round the earth. There are stores of it (dissolved in the oceans, lakes and rivers) in **carbonate** rocks and in fossil **fuels**. A simple diagram of these movements is shown here. Some changes occur quickly, e.g. burning, others very slowly, e.g. formation of rocks.

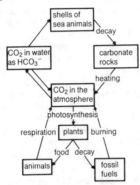

Carbon dating Carbon has a **radioactive isotope** ($^{14}_{6}C$) and the atmospheric **carbon dioxide** contains a small constant proportion of this isotope. All carbon compounds in living tissue contain this carbon-14 isotope in the same constant proportion. When an animal or plant dies, however, the proportion falls at a known rate. It falls because plants no longer take in the isotope through **photosynthesis** and animals no longer take in the isotope by eating plants. All the time the isotope is decaying. By measuring the amount of radioactivity in a dead material it is possible to estimate the age of the material. The technique is used by archaeologists.

Carbonate ($-CO_3$) The carbonate group has a **valency** of two. Metal carbonates occur widely in nature, e.g. **limestone** ($CaCO_3$), dolomite ($CaMg(CO_3)_2$), and malachite ($CuCO_3.Cu(OH)_2$). Calcium carbonate plays a part in the **carbon cycle**.

Carbonates of all **metals** (except **group** I metals) are **insoluble** in water. All carbonates produce **carbon dioxide** when heated *strongly* (K_2CO_3, Na_2CO_3 with difficulty) and when treated with dilute acid. The chemical test of a carbonate is to add acid to the solid compound and to pass the gas produced through **limewater**. A **precipitate**, seen as a milkiness, indicates that the solid was a carbonate.

Catalysis See **Catalyst**.

Catalyst A substance which alters the **rate** of a chemical reaction. The catalyst remains unchanged at the end of the reaction. The process is called *catalysis*.

Transition metals and their compounds are often useful catalysts. Some examples are shown here:

Reaction	Catalyst
Haber process	Iron
Contact process	Vanadium(v) oxide (V_2O_5)
Ammonia → nitric acid	Platinum-rhodium alloy
Hardening fats	Nickel

The chemical processes which go on inside animals

and plants are nearly all dependent on catalysts. These organic catalysts are called **enzymes**.

Cathode The negatively charged pole in a **battery** or **electrolysis cell**. Positively charged **ions (cations)** are attracted to the cathode during electrolysis. They gain electrons at the cathode.

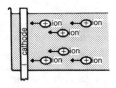

Cation Cations are positively charged **ions**. Most cations are metal ions, e.g. Fe^{2+}, Cr^{3+}, Na^+, Ca^{2+}, Al^{3+} and are attracted to the **cathode** during electrolysis. Three cations not formed from metals are the **hydrogen ion** (H^+) the **oxonium ion** (H_3O^+) and the **ammonium ion** (NH_4^+).

Caustic This term is used to describe substances which burn or corrode organic material, e.g. flesh. It is usually restricted to use with **alkaline** materials, e.g. caustic soda (**sodium hydroxide**), and caustic potash (**potassium hydroxide**).

Celsius scale of temperature (°C) This scale of temperature is based on a 100° range between the melting point of pure **ice** (0°C) and the boiling point of pure water (100°C). Originally called the *centigrade*

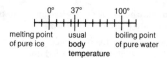

scale, it is now named after Celsius, the man who devised it. One degree Celsius equals one **Kelvin**.

Cell 1. A device for obtaining electrical energy from chemical reactions, often called a **battery**.

2. The single units from which large batteries are made, e.g. the lead-acid **accumulator** used in most cars contains six cells, each of which gives two volts. The total voltage is twelve volts as the cells are connected together in series in the accumulator. The common dry battery is a single cell.

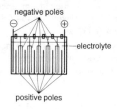

3. The term is also used to describe apparatus and chemicals which are used in electrolysis, e.g. *electrolytic cell*.

Cellulose A **carbohydrate** which is also a **polymer**. It is made up of **glucose monomers**. The formula is $(C_6H_{10}O_5)_n$ and is the material from which cell walls of plants are made. Cellulose is useful to us in two ways. It is used to manufacture paper and **rayon**.

Cement Cement is made by heating **limestone** and clay together and then powdering the product. When water is added and the mixture is allowed to dry a hard substance is produced.

cement + sand + water → mortar
cement + sand + gravel + water → concrete

Centrifuge If it is wished to separate a **solid** from a **liquid**, a centrifuge can be used. The mixture is placed into tubes and put into the machine. The tubes are spun round at very fast speeds and the solid material is forced to the bottom of each tube. Here it forms a compact mass and the liquid can be poured off. Centrifuges are dangerous machines and the lid must always be closed when they are in use.

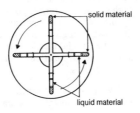

solid material

liquid material

43

Chain Carbon atoms are able to **bond** together in chains and can form *long* molecules. In **polymers**, these chains can be thousands of atoms long. The chains can have branches too. An example of a chain is shown here:

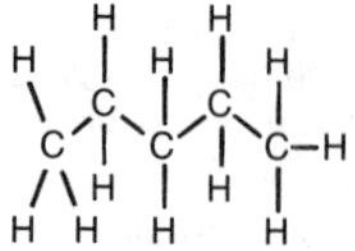

$$C_5H_{12} - \text{pentane}$$

Chain reaction Some **isotopes** are unstable and when they are bombarded with a **neutron** they break up to produce smaller atoms and more neutrons (nuclear **fission**).

$$^{235}_{92}U + \text{neutron} \rightarrow \left\{ \begin{array}{l} \text{atoms or } \textbf{barium} \\ \text{and } \textbf{krypton} \end{array} \right\} + 3 \text{ neutrons}$$

Each time a neutron is used up, three more are produced, each of which can then go on to split up another large unstable atom and produce even more neutrons. These can then go on and more and more neutrons are produced. This is called a chain reaction and is the basis for **nuclear reactions** in nuclear power stations and atom bombs.

Chalk A rock which has been formed from the shells of marine animals. It is mainly **calcium carbonate**

($CaCO_3$). It is a softer rock than **limestone** and its uses are limited by this.

Charcoal Charcoal is made by heating **organic** material (usually wood) in the absence of **air**. In this process, volatile material escapes and the resulting charcoal is mainly composed of **carbon**. Before coal was widely available, charcoal was used to make iron from iron ore. Some charcoal which is called *activated charcoal* is able to absorb small molecules onto its surface. This is used in gas masks, in the **sugar** refining industry and has other uses where coloured impurities need to be removed from materials.

Charles' law The volume of a fixed mass of gas is dependent upon its temperature. If the temperature of a gas is doubled (measured on the **Kelvin** scale) then the volume of the gas will double — *if* the **pressure** of the gas is kept constant.

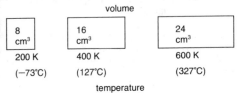

volume

8 cm³	16 cm³	24 cm³
200 K	400 K	600 K
(−73°C)	(127°C)	(327°C)

temperature

Chemical change A change in which one or more chemical substances are changed into *different* ones. Such a change is usually accompanied by the giving

45

out or taking in of **heat energy**. See **Physical change**.

Chemical reaction See **Reaction**.

Chlorides These are **compounds** of chlorine and another element. Metal chlorides are usually **ionic** solids, e.g. sodium chloride (NaCl) and barium chloride ($BaCl_2$).

Nonmetal chlorides are **covalent** compounds and are usually either low boiling point liquids such as tetrachloromethane (CCl_4) or gases such as hydrogen chloride (HCl).

Metal chlorides react with concentrated **acids** to produce **hydrogen chloride** gas:

$$H_2SO_4(l) + NaCl(s) \rightarrow NaHSO_4(s) + HCl(g)$$

The test for chlorides is to mix a solution with silver nitrate solution. If a white **precipitate** forms which then dissolves when **ammonia** solution is added the substance is a chloride.

Chlorination **Chlorine** is added to drinking **water** and to water used in swimming pools to kill dangerous bacteria. The term is also used to describe reactions between chlorine and **hydrocarbons** to produce chlorinated hydrocarbons.

Chlorine (Cl_2) Chlorine is a green **gas** at room temperature. It is a member of the **halogen** group of elements (**group** VII) and is very reactive. It has a

46

choking effect and attacks lung tissue and the throat if it is breathed in. It was used as a chemical weapon in the 1914–18 war.

Chlorine occurs naturally as **chlorides** and sodium chloride is abundant in sea water. Chlorine is extracted by the **electrolysis** of **rock salt** solutions. It is used widely to make **polymers (PVC)**, pesticides **(DDT)**, disinfectants (TCP) and **solvents** (dry cleaning). A solution of chlorine in water is used as a **bleach**.

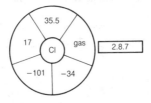

Chlorine is a vigorous **oxidizing agent** which readily reacts with most **elements**. It is made in the laboratory by the oxidation of concentrated hydrochloric acid, e.g.:

$$4HCl(aq) + MnO_2(s) \rightarrow Cl_2(g) + MnCl_2(aq) + 2H_2O(l)$$

Chloroethene (or **vinyl chloride**) The **monomer** from which **poly-(chloroethene)** (PVC) is made.

C_2H_3Cl

It is made by reacting **ethene** (C_2H_4) with **chlorine** (Cl_2)

$$H—Cl +$$

Chlorophyll Chlorophyll is the green pigment in plants which is responsible for the absorption of the sun's energy by the plant. This energy is used in the process of **photosynthesis**. There are two kinds of chlorophyll molecules each of which is a complex molecule which contains a **magnesium** atom.

Chromatography A technique for separating mixtures of **solutes** in a **solution**. One simple method of separating two dyes is to put a spot of the mixture onto the centre of a filter paper (I) and then to add drops of water to the spot. As the water moves out across the paper the dyes begin to separate (II). Eventually two quite separate bands of coloured material will be seen on the paper. The inks in fibre-tipped pens can be separated in this way. The process occurs because some materials cling to the paper more than others, e.g. in the example dye B is more tightly held than A.

Chromium   A **transition metal**. It has important uses in **alloys**, e.g. **stainless steel** and is used in chromium plating on items such as bicycle handlebars, kettles, car bumpers and cutlery. It has a high resistance to corrosion. See **Dichromate (VI) ion**.

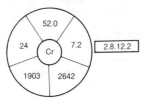

Citric acid A weak organic acid (see **Strengths of acids and bases**) which occurs naturally in *citrus* fruits such as oranges and lemons.

cm³ This is the abbreviation for cubic centimetre and is a unit of **volume** used in scientific work. It is the volume of a cube which has a side of 1 cm. One thousand cm³ are equal to one litre or one cubic decimetre (1 **dm³**).

1 cm

1 cm

Coal A fossilized plant material which was living millions of years ago. It has a very complex chemical structure containing compounds made up of **carbon, hydrogen, oxygen, nitrogen** and **sulphur**. Coal is used as a fuel in power stations, industry and the home and was, before the use of **natural gas**, used as the source of gas (coal gas). About 20% of coal is used to make **coke**.

If coal is heated in the absence of air, coal tar is produced. In the middle of this century coal tar was an important source of **organic** chemicals, e.g. **phenol** from which dyes, drugs and **polymers** were made. Now over 90% of such products come from **petroleum** sources.

As the world has greater reserves of coal than petroleum, attempts are being made to convert coal into petroleum products.

Cobalt A **transition metal**. It is a magnetic element and is used alloyed with iron in the manufacture of magnets. The radioactive **isotope** $_{27}^{60}$Co emits **gamma rays** and it is used in the treatment of cancers. Cobalt(II) chloride is blue when **anhydrous** and pink when **hydrated**, and so the compound is used as a test for the presence of water.

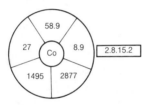

Coke Coke is the material left behind when the **volatile** compounds have been removed from **coal**. It contains over 80% carbon. It was made during the production of coal gas but now is made in *coke ovens* from special *coking coal*. Coke is used to make **carbon monoxide** in the **blast furnace** and is an important (smokeless) fuel.

Combining power See **Valency**.

Combustion In combustion (or burning) reactions,

a substance combines with a **gas**. Heat and light (i.e. **flame**) usually accompany combustion reactions. Most combustions involve **oxygen**, e.g.:

$$2H_2(g) + O_2(g) \rightarrow 2\,H_2O(g)$$

Complex ions A **cation** (e.g. metal ions or hydrogen ions) which is bonded to one or more small molecules by a **coordinate bond.** The simplest examples are the ammonium ion and the oxonium ion:

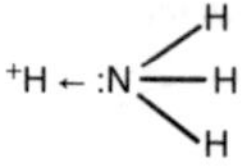

NH_4^+

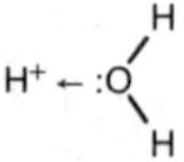

H_3O^+

Here the **lone pair of electrons** on the nitrogen and oxygen atoms forms the bond.

Transition metal ions form a large number of complex ions because they can easily accept the donated **electron** pairs.

Compound A compound is a pure substance which is made up of two or more elements chemically bonded together. The **properties** of a compound are quite different from the properties of the elements bonded together within it. Compounds may contain **ionic** or **covalent** bonding. Examples: methane CH_4, water H_2O, sodium chloride NaCl.

See **Mixture**.

Concentrated A concentrated solution is one which contains a relatively large concentration of solute. If you want to increase the concentration of a solution, this can be done by either adding more solute or by removing solvent (e.g. by **distillation**) from the solution. See **Dilute**.

Concentration The concentration of a **solution** is a measure of how much **solute** is **dissolved** in the solution. It is usually expressed in terms of how much substance is present in a given volume of the solution. This can either be in terms of **mass**, e.g. grams, or in terms of how many particles, e.g. **moles**. Examples of concentrations and how they can be written are: for a solution containing 170 grams of silver nitrate (1 mole) in one cubic decimetre (**dm^3**) of solution (see **molarity**):

170 g/dm^3, or 1 mol/dm^3 or 1 mol/l or 1 M.

Condensation **1.** The change from the **gas** or **vapour** state to the **liquid** state, e.g. water condensing onto a cold window pane.

2. A reaction in which two or more molecules react to produce a larger molecule and a small molecule such

53

as water. This is one method of producing **polymers**, e.g. **nylon**.

Conductor An *electrical* conductor is a substance which will allow an electric **current** to flow through it. **Metals**, solutions which contain **ions** and molten **ionic** compounds are conductors; all other substances are **insulators**. Graphite (a form of carbon) is a notable exception. Certain **metalloids** are **semiconductors**.

Examples of conductors are:

Sodium (Na). Conduction by **electrons**.	Molten copper(II) chloride (CuCls(l)). Zinc chloride solution (ZnCl$_2$(aq)). Conduction by movement of **ions**.

The term conductor is also applied to substances which allow **heat energy** to flow through them. Metals are also good conductors of heat.

Conservation of energy and mass In a chemical reaction the **mass** of the reacting substances and the mass of the products are equal. Similarly:

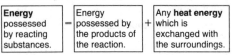

Energy possessed by reacting substances.	=	Energy possessed by the products of the reaction.	+	Any **heat energy** which is exchanged with the surroundings.

In other words mass and energy cannot be created or destroyed in a chemical reaction. They are *conserved*.

These relationships are only approximations. The work of Einstein showed that mass and energy can be

54

converted into each other. This is important in **nuclear reactions** but in ordinary chemical reactions the relationship holds good.

Contact process This process is used to make almost all the **sulphuric acid** produced today. The flow chart below shows the steps:

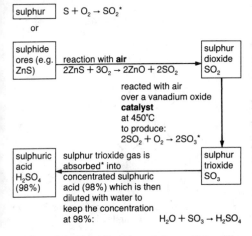

sulphur	$S + O_2 \rightarrow SO_2^*$

or

| sulphide ores (e.g. ZnS) | reaction with **air** $2ZnS + 3O_2 \rightarrow 2ZnO + 2SO_2$ | | sulphur dioxide SO_2 |

reacted with air over a vanadium oxide **catalyst** at 450°C to produce: $2SO_2 + O_2 \rightarrow 2SO_3^*$

| sulphuric acid H_2SO_4 (98%) | sulphur trioxide gas is absorbed* into concentrated sulphuric acid (98%) which is then diluted with water to keep the concentration at 98%: $H_2O + SO_3 \rightarrow H_2SO_4$ | | sulphur trioxide SO_3 |

Most sulphuric acid is now made from **sulphur**. The reactions marked * are **exothermic** and so the process

55

is relatively cheap to run because little heat energy has to be supplied.

Coordinate bond   A **covalent bond** in which *both* the **electrons** in the bond have come from the *same* **atom**. Electrons have been *donated* by one atom rather than one donated from each atom. The bond is also known as a dative bond. Examples are:

$$H-\underset{\underset{\displaystyle H}{|}}{\overset{\overset{\displaystyle H}{|}}{N}}: \rightarrow H^+ \rightarrow NH_4^+$$

$$\rightarrow [Cu(H_2O)_6]^{2+}$$

In coordinate bonding, the donor atom uses a **lone pair of electrons**. See **Complex ion**.

Copper   A **transition element** which plays an important part in our lives. It is a vital trace element in our bodies — we need between one and two milligrammes per day.

Copper is an unreactive metal, coming low down in the **electrochemical series**. It will not liberate **hydrogen** from dilute **acids**.

It can be purified by using the impure metal as the **anode** with copper(II) sulphate as the **electrolyte** and a small pure copper cathode in an **electrolysis cell**.

Copper metal is used in plumbing and for electrical wiring. (Copper is an excellent **conductor**.) It is extracted commercially from its **sulphide** ores.

| Chalcopyrite | $(CuFeS_2)$ |
| Bornite | (Cu_5FeS_4) |

and its carbonate ores:

| Malachite | $(CuCO_3.Cu(OH)_2)$ |
| Azurite | $(2CuCO_3.Cu(OH)_2)$ |

Copper is also found as the free element, an indication of its lack of reactivity.

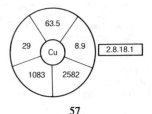

Copper compounds

Copper(II) chloride $CuCl_2.2H_2O$	Copper (II) chloride is used to give fireworks a green colour and in the removal of **sulphur** from **petroleum**.
Copper(II) hydroxide $Cu(OH)_2$	Copper (II) hydroxide is used in the manufacture of **rayon**, in dyeing textiles and as a blue pigment.
Copper(II) carbonate $CuCO_3$	Copper (II) carbonate forms part of the ores malachite and azurite — both basic copper carbonates. Another basic carbonate — verdigris — is formed as a surface layer on copper which is exposed to the **atmosphere**.
Copper (II) sulphate $CuSO_4.5H_2O$	Copper sulphate forms blue crystals and can be made by treating copper compounds (oxide, carbonate, hydroxide) with dilute sulphuric acid. The **anhydrous** salt is a white powder which turns blue when water is added:

$$CuSO_4 + 5H_2O \rightarrow CuSO_4.5H_2O$$
white $\qquad\qquad$ blue

This reaction is often used as a test for the presence of water. The compound is used in wood preservatives and fungicides, e.g. Bordeaux mixture.

Corrosion When a substance corrodes it is chemically eaten away. Good examples are the weathering of **limestone** buildings by rainwaters which contain dissolved acids; **rusting** of **steel**. Corrosion begins at the surface and often a surface layer protects the rest of the material, e.g. the **oxide** coating on **aluminium**. Rusting is not like this, the rusting process goes through the steel until it is all corroded. That is why it is so damaging and expensive.

Coulomb *Electric charge* is measured in coulombs. It is the product of the **current (units = ampère)** and the time (units = **seconds**), e.g. if a current of 0.6 amps flows for ten seconds, six coulombs of electricity will have passed through the circuit.

Covalent bonds These **bonds** are formed by two **atoms** coming together and sharing their **electrons**, e.g. **hydrogen** atoms have one electron each (•). The atoms form a **molecule** of hydrogen (H_2) by the two electrons forming a bond.

Thus each 'shared pair' of electrons produces one covalent bond.

hydrogen atoms (H)

hydrogen molecule (H_2)

hydrogen chloride

Covalent bonds are usually found in compounds which only contain **nonmetallic** elements, e.g.

Carbon dioxide	CO_2
Hydrogen chloride	HCl
Nitrogen dioxide	NO_2
Ammonia	NH_3
Phosphorus(v) oxide	P_2O_5

and also in elements with an **atomicity** of two or more, e.g.

Oxygen	O_2	Hydrogen	H_2
Bromine	Br_2	Chlorine	Cl_2
Nitrogen	N_2	Sulphur	S_8
Ozone	O_3	Phosphorus	P_4

Covalent bonds which contain two electrons i.e. one 'shared pair' are termed *single* bonds. Many molecules contain **double** or **triple** bonds. Covalent bonds are not as strong as **ionic bonds**.

Covalent compounds These are compounds which contain **covalent bonds**. They tend to be non-**conductors**. Their **melting** and **boiling points** are low.

Cracking This is the process in which *large* **hydrocarbon molecules** are broken up into *small* molecules. This occurs in the **petroleum** refining industry. Small molecules are more valuable than large ones because they are the starting materials needed in the production of such things as **polymers**, e.g. **poly(ethene)** and **poly(propene)**, and fuels, e.g. **petrol** and **diesel fuel**.

Different products are formed under different conditions:

Conditions	Process name	Products
High temperature.	Thermal cracking.	**Unsaturated** and **saturated** molecules are produced.
High temperature in the presence of steam.	Steam cracking.	
High temperature in the presence of a catalyst.	Catalytic cracking.	
High temperature and pressure in the presence of hydrogen.	Hydro-cracking.	The product is totally saturated.

Cryolite In the extraction of **aluminium, bauxite** (Al_2O_3) is dissolved in molten cryolite (Na_3AlF_6) at 900°C and the mixture is then electrolysed. Aluminium is produced at the **cathode**.

Crystal Solid substances have their atoms (molecules or ions) arranged in a set geometrical 3-dimensional pattern. The shape of these atomic arrangements (crystal **lattices**) determines the shapes that the solid has when it undergoes **crystallization**.

Some examples of different crystal arrangements:

cubic   octahedral   tetrahedral

Crystallization This is the process in which **crystals** are formed. In nature, crystals are produced when molten rocks cool down and solidify. In laboratory experiments, crystals are usually produced from a **solution**. There are two main methods:

  1) A solution is left at **room temperature**. Slow **evaporation** of the **solvent** takes place and crystals are left behind.

  2) A **supersaturated solution** is made above room temperature. As this is cooled down, crystals are produced.

Current Electric current is the movement of **electrons** through a **conductor**. It is measured in **ampères** (amps). Current is said to move from positive terminal to negative terminal. The electrons, however, travel from the negative terminal to the positive. This difference is because the electron carries a negative electric charge.

Cyanide Compounds which contain the $-CN$ group of atoms are known as cyanides. Potassium cyanide (K^+CN^-) and hydrogen cyanide ($H-C \equiv N$) are the best known examples. They are very poisonous.

Cyanides prevent the flow of **energy** within the body and so are effective very quickly.

DC (or **direct current**) The type of electrical current produced from a simple cell or battery. Only DC can be used in electrolysis.

DDT (or **DichloroDiphenylTrichloroethane**) A

widely used pesticide which has been successful in the control of diseases such as *malaria*. It is, however, harmful to animal life as some animals can concentrate it in their bodies where it acts as a **poison**. For this reason some countries restrict its use.

Decant To pour off. If a **precipitate** is allowed to settle, the **liquid** remaining can then be decanted by pouring it carefully away without disturbing the solid.

Decomposition The break up of compounds into simpler compounds or into elements. Usually heat is needed (thermal decomposition), e.g.

$$2Cu(NO_3)_2(s) \rightarrow 2CuO(s) + 4NO_2(g) + O_2(g)$$
$$2HgO(s) \rightarrow 2Hg(l) + O_2(g);$$
$$CaCO_3(s) \rightarrow CaO(s) + CO_2(g)$$

Decrepitation When some **crystals** are heated a sharp crackling noise is heard. This is the sound of the crystals breaking up. Lead(II) nitrate (Pb(NO$_3$)$_2$) is an example of a salt which undergoes decrepitation.

Dehydrating agent A dehydrating agent is a substance which is used to remove water from other substances. Such substances always have an attraction for water. They can be substances which dissolve in water, e.g.:

Concentrated sulphuric acid H_2SO_4	Sodium hydroxide NaOH

or substances which react with water, e.g.:

Calcium oxide CaO	this forms the hydroxide $\rightarrow Ca(OH)_2$

A third kind of dehydrating agent is the **anhydrous** salt which absorbs water, e.g:

Calcium chloride $CaCl_2$ $\rightarrow CaCl_2.2H_2O$	Sodium sulphate Na_2SO_4 $\rightarrow Na_2SO_4.10H_2O$

Dehydrating agents are used in **desiccators**.

Deliquescence Deliquescent substances are able to take in water from the **atmosphere**. They can take in so much water that they are able to form a solution — unlike **hygroscopic** substances.

Some common examples are:

Iron(III) chloride $FeCl_3$	Copper(II) nitrate $Cu(NO_3)_2$	Calcium chloride $CaCl_2$

Some deliquescent substances are used in **desiccators**.

Delocalized electrons In normal **covalent bonds, electrons** are found between two **atoms**, e.g. a carbon-carbon **single bond** consists of two electrons:

$$C : C$$

and a double bond contains four electrons:

$$C :: C$$

These electrons are located in between the atoms. They are *localized*. In some substances, however, there are some electrons which are free to move amongst the atoms. These electrons are *delocalized*. **Graphite** contains delocalized electrons *within* the hexagonal layers of atoms. This is why graphite can conduct electricity along the layers *but not* between them. **Metals** have delocalized electrons too and this explains their being good **conductors**.

Density The density of a substance is a measure of how much space a certain amount of the substance takes up. The smaller the space into which the **mass** is concentrated, the *greater* is the density. The density of a substance, e.g. lead, is determined by the density of the individual **atoms** *and* by how closely together the atoms are arranged in the **crystal lattice**. Generally speaking, the more particles that an atom contains, the denser it will be.

Density is calculated by dividing a substance's mass

by its volume. The units are usually either g/cm³ or kg/m³.

Desiccator Desiccators are used to store substances which need a dry atmosphere, e.g. **deliquescent** or **hygroscopic** substances.

A desiccator is airtight and contains a **dehydrating agent** which makes the air dry.

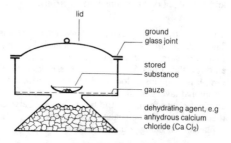

Detergent A cleaning agent. It does two things when used in washing, e.g. clothes or dishes. It reduces the surface tension of the water and so allows the water to wet things more thoroughly. It also acts to bring together the water and (normally insoluble) fat, **oil** or grease into an emulsion. If the presence of a detergent is accompanied by rapid movement, e.g. in a washing machine, then particles of dirt and grit are removed and the article is cleaned.

Detergents form an emulsion because one part of

their molecule is **ionic** and is attracted to water and because the other part (the **hydrocarbon** chain) is attracted to the oil molecules. Detergents, thus, hold the two together. An example of a synthetic detergent molecule:

$$
\begin{array}{c}
\text{H H H H H H H H H H H H H H} \quad\quad \text{O} \\
\mid\ \mid\ \mid\ \mid\ \mid\ \mid\ \mid\ \mid\ \mid\ \mid\ \mid\ \mid\ \mid\ \mid \quad\quad \parallel \\
\text{H}-\text{C}-\text{C}-\text{C}-\text{C}-\text{C}-\text{C}-\text{C}-\text{C}-\text{C}-\text{C}-\text{C}-\text{C}-\text{C}-\text{C}-\text{O}-\text{S}-\text{O}^-\ \text{Na}^+ \\
\mid\ \mid\ \mid\ \mid\ \mid\ \mid\ \mid\ \mid\ \mid\ \mid\ \mid\ \mid\ \mid\ \mid \quad\quad \parallel \\
\text{H H H H H H H H H H H H H H} \quad\quad \text{O}
\end{array}
$$

See **Soap**, **Hydrophilic** and **Hydrophobic**.

Deuterium An **isotope** of hydrogen which has a **neutron** and a **proton** in the **nucleus**. Its symbol is written:

$$^2_1\text{H} \text{ or } ^2_1\text{D}.$$

The isotope occurs naturally, making up 0.015% of hydrogen. Because its density is twice that of normal hydrogen it is easily separated.

Deuterium oxide is known as *heavy water* and its formula is written D_2O.

Di- A prefix which means *two*, e.g.:

Carbon *di*oxide CO_2 — two atoms of oxygen.
*Di*atomic molecule — two atoms in the molecule.

Diamond An **allotrope** of **carbon**. It occurs naturally — mainly in South Africa and the Soviet Union and is a valuable mineral. It is the hardest naturally occurring substance known and is used in

67

drill tips and saw blades. Diamond is prized as a gemstone because of its rarity and the sparkle it produces. The carbon **atoms** are arranged in the **crystal lattice** in a **tetrahedral** formation and are covalently bonded to each other.

Diatomic molecule A **molecule** which contains two atoms is said to be **di**atomic, e.g.

N_2 O_2 H_2 Cl_2 F_2 Br_2 I_2.

Dibasic acid An **acid** which contains two replaceable **hydrogen** atoms per molecule is called a dibasic acid.

Some examples are shown. With such acids, two kinds of salt can be formed — the *normal* salt, in which *both* hydrogen atoms are replaced, and the **acid salt** in which only one hydrogen has been replaced. Examples of these are also shown.

Acid	Salts	
	acid	*normal*
H_2SO_4 sulphuric acid	$NaHSO_4$ sodium hydrogen-sulphate	Na_2SO_4 sodium sulphate
H_2CO_3 carbonic acid	$NaHCO_3$ sodium hydrogen-carbonate	Na_2CO_3 sodium carbonate

Dichromate(VI) ion ($Cr_2O_7{}^{2-}$) This ion contains **chromium** and **oxygen** atoms and has a valency of two. Its formula is $Cr_2O_7{}^{2-}$ and it is usually used as the bright orange potassium or ammonium salts:

$$K_2Cr_2O_7 \quad (NH_4)_2Cr_2O_7$$

Dichromate ion is an **oxidizing agent** which is reduced to chromium(III) ions Cr^{3+}.

Diesel fuel A fuel/air mixture is injected into diesel engines where it is compressed. The temperature produced in the compression causes the fuel/air mixture to explode. The fuel is produced from **petroleum** and is made up mainly of **alkanes** in the boiling range 200°C–350°C. It is often sold as 'DERV' fuel: *D*iesel, *E*ngine, *R*oad, *V*ehicle.

Diffusion If two gases (e.g. air and nitrogen dioxide) are at first separated and then allowed to mix, they mix together *completely*. This happens because the molecules in the gases are moving about randomly at high speed because of the **thermal energy** they possess.

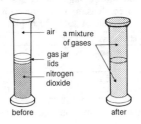

Diffusion occurs at a faster rate when the temperature is raised. The *less* dense a gas is, the *greater* is its rate of diffusion.

Diffusion also occurs in liquids, solutions and solids.

Dilute A dilute **solution** is one which contains a relatively low **concentration** of **solute**. If you want to make a solution *more* dilute it is usual to add more solvent to it. Dilute acids usually have concentrations of 2 **mol/dm³** *or less*.

Dimer A **molecule** which is made up of two identical molecules (**monomers**) which are bonded together. Example:

N_2O_4 is a dimer of NO_2

Disaccharide A **sugar** molecule which is made up of two **monosaccharide** sugar molecules which have undergone a **condensation** reaction with the elimination of a molecule of water. Examples:

Monosaccharides		Disaccharide
glucose + glucose $C_6H_{12}O_6$ $C_6H_{12}O_6$		maltose $C_{12}H_{22}O_{11}$
glucose + fructose $C_6H_{12}O_6$ $C_6H_{12}O_6$		sucrose $C_{12}H_{22}O_{11}$

Disinfectant A disinfectant is a substance which is

capable of destroying harmful bacteria. They are often based on the compound **phenol**; two common household examples are shown here:

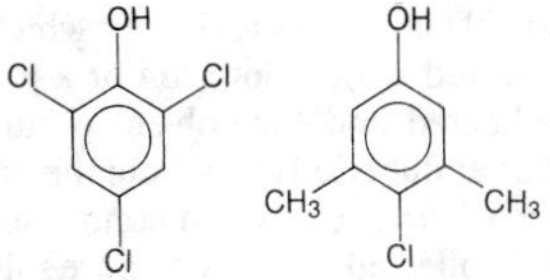

used in	used in
T.C.P.	*Dettol*

Displacement reactions   In a displacement reaction a less reactive element is displaced by a more reactive one. These are **redox** reactions.

Example:

$$CuSO_4(aq) + Zn(s) \rightarrow ZnSO_4(aq) + Cu(s)$$
$$2NaBr(aq) + Cl_2(g) \rightarrow 2NaCl(aq) + Br_2(aq)$$

Dissociation   When a compound dissociates, it breaks up into smaller simpler molecules or ions. The dissociation is sometimes **reversible**.

$$NH_4Cl(s) \overset{heat}{\underset{cool}{\rightleftharpoons}} NH_3(g) + HCl(g)$$

| ammonium chloride | ammonia | hydrogen chloride |

$$HCl(aq) \rightleftharpoons H^+(aq) + Cl^-(aq)$$

71

Dissolve When a **solute** dissolves in a **solvent** to form a **solution** it is dispersed throughout the whole volume of the solvent. The structure of the solute is broken up.

Distillation This is the process by which a **solvent** can be recovered from a **solution** or a **mixture**. The solution is heated and the solvent is turned into a vapour. The vapour is led away from the hot flask into a condenser where it cools and condenses to a pure liquid, and is collected. In this way pure solvent can be recovered. A typical set of apparatus is shown.

See **Fractional distillation**.

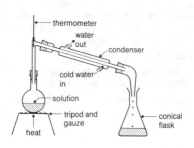

Distilled water Tap water and rain water are not **pure**. They contain dissolved **salts** and **gases**. Water is often distilled to increase its purity. Most of the salts are left behind but the water still may contain dissolved gases. The presence of **carbon dioxide** reduces the **pH** of the water considerably.

dm³ This is the symbol for cubic decimetre, 1000 **cm³ = dm³**.

DNA Deoxyribonucleic acid is the very complex chemical which controls the order in which **amino acids** combine to produce **proteins**.

Double bond A chemical **bond** which contains *two shared pairs* of **electrons** is a double bond. Carbon-carbon double bonds are found in **alkene** molecules. The presence of the extra electrons makes the alkenes very reactive. Such **compounds** which contain double (and **triple**) bonds are said to be **unsaturated** compounds.

Compounds and elements containing double bonds:

Oxygen $O = O$

Ethene

$$\underset{H}{\overset{H}{\diagdown}} C = C \underset{H}{\overset{H}{\diagup}}$$

Carbon dioxide $O = C = O$

Double decomposition (or **Precipitation reactions**) This process occurs when ionic substances react in solution. The product is an insoluble solid. Example:

$$CaCl_2(aq) + Na_2CO_3(aq) \rightarrow CaCO_3(s) + 2NaCl(aq)$$
$$or \; Ca^{2+}(aq) + CO_3^{2-}(aq) \rightarrow CaCO_3(s)$$

73

This is a method of preparing insoluble **salts**.

Dry cell This is a **battery** in which the electrolyte is in the form of a paste and so cannot spill from the battery.

The only common *wet* battery used today is the lead-acid **accumulator**.

Dry ice Solid Carbon dioxide. This material **sublimes** and hence the name *dry*. It is used for keeping things cold (sublimation point = $-40°C$) and in the theatre for producing smoke and clouds on stage.

Ductile A ductile substance can easily be drawn into wires, e.g. **copper** is a very ductile **metal**. The term is usually used only about metals. Ductile metals have large **crystals** within them. See **Annealing** and **Malleable**.

Duralumin An **alloy** of **aluminium**. It contains **copper** (4%) and traces of **manganese, magnesium** and **silicon**. The alloy is much *harder* and *stronger* than the pure metal. It is used for the production of aircraft frames and cooking utensils.

Dyes Chemicals which can be mixed or reacted with materials to make a coloured product, e.g. fabrics, plastics. The colour is produced because the dye absorbs some of the light which falls upon it and radiates the rest.

Dynamite This explosive was invented by Alfred Nobel. He discovered that if the unstable explosive nitroglycerine was absorbed into a clay called kieselguhr the result was a *stick* of explosive which was safe until it was detonated.

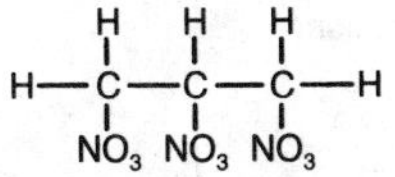

Nitroglycerine

Efflorescence The process in which crystals lose part of their **water of crystallization** when they stand in the air. They go powdery on their surfaces. A well-known example of this process is sodium carbonate (washing soda):

$$Na_2CO_3.10H_2O(s) \rightarrow Na_2CO_3(s) + 10H_2O(g)$$

Electric current or **Electricity** See **Current**.

Electrochemical series **Elements** differ in their reactivity. When a metal is dipped into a **solution** of one of its salts, e.g. zinc in zinc sulphate solution, the following reaction occurs:

zinc → zinc ions + 2 electrons
$$Zn(s) \rightarrow Zn^{2+}(aq) + 2e^-$$

The ions go into solution and the **electrons** cling to the metal. If the metal and solution are made part of a circuit the electrons will flow round the circuit. The *more* reactive a metal is the greater the force with which the electrons move. See table over page.

K	potassium	↑ reactivity *increases*
Na	sodium	
Ca	calcium	
Mg	magnesium	
Al	aluminium	
Zn	zinc	
Fe	iron	
Pb	lead	
H	hydrogen	
Cu	copper	
Ag	silver	

When metals are put in order of the force they produce it is known as the 'electrochemical series': The series is useful because it allows us to compare reactivities and hence predict reactions, e.g.:

zinc *will* remove the oxygen from copper (II) oxide, but

copper *will not* remove oxygen from zinc oxide; hydrogen *will* reduce copper (II) oxide *but not* zinc oxide

copper *will not* react with acids to release hydrogen.

Electrode   An electrode is a **conductor** which dips into an **electrolyte** and allows the **current (electrons)** to flow to and from the electrolyte. **Copper, carbon** and **platinum** are used in **electrolysis** as electrodes. In electrolysis chemical reactions occur *at* the electrodes. See **Anode, Cathode**.

Electrolysis   When a *direct* electric current (**DC**) is passed through a liquid which contains ions (an

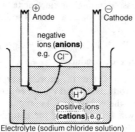

Electrolyte (sodium chloride solution)

electrolyte) **chemical changes** occur at the two
electrodes.

at the anode	at the cathode
$Cl^- \rightarrow Cl + e^-$	$H^+ + e^- \rightarrow H$
$2Cl \rightarrow Cl_2$	$2H \rightarrow H_2$
chlorine is given off	hydrogen is given off

$$2[H^+ + Cl^-] \rightarrow H_2 + Cl_2$$

Electrolysis is used to extract metals and nonmetals
from compounds. Aluminium, sodium and copper are
produced in this way, as is chlorine.

Electrolyte An electrolyte is either a molten ionic
compound:

$$NaCl(l) \quad PbI_2(l)$$

or a solution which contains ions:

$$HCl(aq) \quad NaOH(aq) \quad KI(aq) \quad CuSO_4(aq)$$

Chemical changes take place when a direct electric **current** is passed through an electrolyte (**electrolysis**).

With molten compounds such as sodium chloride the changes are simple:

$$2Na^+Cl^-(l) \rightarrow 2Na(l) + Cl_2(g)$$

The elements are produced.

With ionic solutions, the processes are more complex. The products depend on the **concentration** of the solution, the voltage which is applied, the type of **electrode** which is used, and the nature of the ions present in the solution. Example:

Electrolyte		Anode	Reaction
	(a)	carbon or platinum	$4OH^-(aq) \rightarrow$ $2H_2O(l) + O_2(g) + 4e^-$
$CuSO_4(aq)$			
	(b)	copper	$Cu(s) \rightarrow Cu^{2+}(aq) + 2e^-$

In one case (a) oxygen is released; in the other (b) the copper anode dissolves. Process (b) is used in the purification of copper.

Electron　Electrons are very small particles with an extremely tiny mass (about $\frac{1}{1840}$ of that of the **proton**). They have a negative charge and move round the nuclei of atoms. They are given the symbol e^-.

When atoms lose or gain electrons they form ions:

gain → negative **anions**, e.g. Cl^- O^{2-} Br^- S^{2-}
lose → positive **cations**, e.g. Na^+ H^+ Fe^{2+} Al^{3+}

An electric **current** is electrons moving through a conductor. The way in which electrons are located in an atom is known as the **electronic configuration**.

Electronegativity The electronegativity of an atom is the power that it has to attract **electrons** to itself in a chemical bond.

Li	Be	B	C	N	O	F

Li Be B C N O F
→ electronegativity increases *across* a period and *up* a group

Na	Cl
K	Br
Rb	I
Cs	At

It follows that **flourine** has the *most* electronegative atoms.

Electronic configuration (arrangement) **Electrons** move around the **nucleus** of an **atom**. The movement takes place in clearly-defined regions (orbitals) in the atoms. Electrons are pictured as being arranged in orbits, or what are called **shells**, each of which can contain a *maximum* number of electrons.

Shell number:	①	②	③	④	⑤
Maximum number of electrons:	2	8	18	32	50

As we go across the **periodic table** the number of electrons increases by one per atom as the **atomic number** increases. The shells are filled up in turn — ① first, then ②, etc. — so that the electrons go into the region of the atom which has the lowest **energy**. For some elements described in this book, the electronic configuration is given, e.g. sodium: 2.8.1, i.e., the ① and ② shells are full and the ③ shell is beginning to fill.

Different configuration patterns are found in the periodic table, e.g. **group** I elements always have an electronic configuration ending in 1:

Lithium   2·1
Sodium   2·8·1

The **halogens** Group VII have a different pattern; which in fact ends in 7 (the group number).

Fluorine   2·7
Chlorine   2·8·7
Bromine   2·8·18·7

Electroplating If a metallic object is made the **cathode** in an **electrolysis** where the **electrolyte** contains metal **ions**, the object can be coated (plated) with a very thin layer of metal. Conditions — **current**, electrolype **concentration** and **temperature** must be carefully controlled.

Electroplating is usually carried out for either (a) protection, e.g. chromium plating on handlebars and motor car bumpers and tin plating on cans, or (b) decoration, e.g. silver plating cutlery and ornaments, and gold plating jewellery, etc.

Electrovalent bond See **Ionic bond**.

Element A **pure** substance which cannot be broken

down into anything simpler by chemical means is said to be an element. There are 104 elements known to man. Most of these occur naturally on the earth but several have been made in laboratories by **nuclear reactions.** All elements have a unique number of **protons** in their atoms. There are three classes of element:

metals	nonmetals	metalloids
e.g. **iron**	e.g. **oxygen**	e.g. **arsenic**

Empirical formula The empirical formula of a compound shows the atoms that are present in the molecule in their simplest ratio:

Compound	Molecular formula	Empirical formula
Ethene	C_2H_4	CH_2
Butane	C_4H_{10}	C_2H_5
Propane	*C_3H_8	C_3H_8
Ethanoic acid	$C_2H_4O_2$	CH_2O

* In some examples the empirical and molecular formulae are the same.

See **Molecular formula** and **Percentage composition**

Endothermic reaction A reaction in which **heat energy** is taken in from the surroundings when a reaction occurs.

There is either a fall in temperature when the

81

reaction occurs, e.g. when sodium nitrate dissolves in water, e.g.:

$$NaNO_3(s) \rightarrow NaNO_3(aq)$$

or heat energy has to be continually supplied to make a reaction occur, e.g.:

$$CaCO_3(s) \rightarrow CaO(s) + CO_2(g)$$

In endothermic reactions, there is *more* **energy** in the bonds of the products than there was in the bonds of the reactants.

End point The point at which a reaction is complete.

The end point of a **titration** is when *all* of one of the reactants has been used up. The end point can be identified by an **indicator** or an instrument such as a **pH** meter or a conductivity meter. Thus the end point is when the indicator changes colour.

Energy The 'ability to do useful work'. Energy is *locked* inside the nuclei of atoms. Sometimes this can be released — see **Nuclear reactions**. Energy is also found in the **bonds** between atoms. When chemicals react bonds break and new ones form. In this breaking and forming, energy can be *absorbed* or *released* as heat energy — see **Energy change**. But in batteries it is released as electrical energy. Energy is usually measured in **joules** or **kilojoules**. See **Enthalpy**.

Energy change When chemical reactions occur, heat energy can be released from the reactants, − **exothermic**; or can be taken in by the reactants, − **endothermic**.

Energy changes are measured in **joules** or **kilojoules** and refer to a particular amount of substance − usually a **mole**. So, the energy change for the **Haber process** is − 92 kilojoules per **mole** (−92 kJ/mol).

Enthalpy (*H***)** The amount of **heat energy** possessed by a chemical substance. It is given the symbol H. It cannot be measured but, when chemicals react the difference in the heat energy between the reactants and products can be measured by using a **thermometer**. This difference is the enthalpy change and is given the symbol ΔH. ΔH is negative for an exothermic reaction and positive for an endothermic reaction.

Enzyme Enzymes are usually **protein** molecules. They are **catalysts** which are found in living tissue. They allow complicated biochemical reactions to occur at low temperatures and pressures inside the body. It is thought that an enzyme is a particular shape and this shape allows two reacting molecules to come close together on the enzyme. This makes the reaction easy to carry out, e.g.:

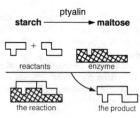

Equation In chemistry, a reaction can be described by an equation. The equation can be a *word* equation:

$$\text{hydrogen} + \text{oxygen} \rightarrow \text{water}$$

or it can be a *formula* (or *symbol*) equation:

$$H_2 + O_2 \rightarrow H_2O$$

An equation is said to be *balanced* if there are the same number of each kind of atom on each side of the equation, e.g.:

$$2H_2 + O_2 \rightarrow 2H_2O$$

is a balanced equation.

Two molecules of hydrogen react with one molecule of oxygen to produce two molecules of water. It also tells us that 2 moles or 4g of hydrogen react with 1 mole or 32g of oxygen to produce 2 moles or 36g of water. Since relative atomic mass H = 1, O = 16:

$$2H_2 + O_2 \rightarrow 2H_2O$$
$$2(2) + 32 = 2(2 + 16) = 36$$

Sometimes we use **ionic equations**:

$$Cu^{2+} + Zn \rightarrow Cu + Zn^{2+}$$

State symbols are usually added after the formula to denote which **state of matter** the substance is in, e.g. gaseous hydrogen and oxygen react to produce liquid water:

$$2H_2(g) + O_2(g) \rightarrow 2H_2O(l)$$

Equilibrium In a **reversible reaction** the reaction proceeds in both directions, e.g.:

$$3Fe(s) + 4H_2O(g) \rightleftharpoons Fe_3O_4(s) + 4H_2(g)$$

The reactants (iron and steam) produce the oxide and hydrogen. As soon as the products are made, they begin to react to reform steam and iron. There eventually comes a time when the **rate** of the forward $\rightarrow$ reaction equals the rate of the reverse $\leftarrow$ reaction. At this point the proportion of substances present is constant. There seems to be no reaction taking place. There is an *equilibrium* between the forward and reverse reactions.

It is important to realize that such an equilibrium is dynamic — the reactions are occurring but because they are taking place at the same rate no charge can be seen.

Many important reactions involve equilibria.

In the **Haber process**, with the usual operating conditions *less than 20%* of the **hydrogen** and **nitrogen** are converted into **ammonia**. This is the equilibrium **yield** of ammonia in the reaction.

Ester A compound formed between an **alcohol** and a carboxylic **acid**.

A **strong acid catalyst** is usually needed for the reaction (see diagram on opposite page).

Esters often have sweet, *fruity* smells and are used in perfumes, flavourings and essences.

The reaction between acid and alcohol is known as *esterification*.

Ethane An **alkane**. It is a colourless, flammable gas which is found in **natural gas** to a small extent.

C_2H_6

H H
| |
H—C—C—H
| |
H H

Ethanoic acid (or **acetic acid**) A colourless liquid (**melting point** 16°C, **boiling point** 118°C).
It is a carboxylic acid which is made industrially by

ethanoic acid · ethanol

ethyl ethanoate ($CH_3COOC_2H_5$) · water

the oxidation of **naphtha**. It is used in the production of important fabrics such as *Tricel*.

Ethanoic acid is present in **vinegar** (about a 4% aqueous solution).

Although it is a **weak acid**, in a concentrated form it can cause severe skin burns.

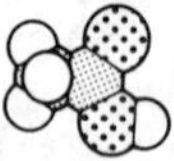

CH₃CO₂H

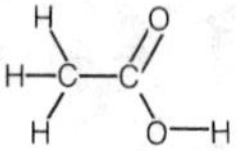

Ethanol A colourless, flammable **alcohol** whose **boiling point** is 78°C. Ethanol is the alcohol contained in alcoholic drinks. These are made by **fermentation**, but in the chemical industry it is made by the hydration of **ethene**:

$$C_2H_4(g) + H_2O(l) \rightarrow C_2H_5OH(l)$$

C₂H₅OH

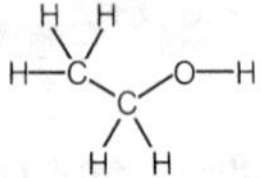

Ethanol is used in industry as a solvent. It is also used in **methylated spirits.** It can be oxidized to ethanal and **ethanoic acid.**

Ethene (formerly **ethylene**) Ethene is a gaseous **alkene** which is produced by **cracking alkanes** (e.g. **naphtha**).

Because it is an **unsaturated** compound it is reactive. It is a very important chemical being used to make **plastics** such as **poly(ethene)**.

C_2H_4

Ethene can add molecules across its **double bond**, e.g. to produce **ethanol**:

$$C_2H_4 + H_2O \rightarrow C_2H_5OH$$

$$C_2H_4 + Br_2 \rightarrow C_2H_4Br_2$$

These are addition reactions.

Ethers Ethers are **organic compounds** which have an oxygen atom which is bonded to two carbon atoms.

$$C_4H_{10}O$$

One of the most important is diethyl ether (ethoxyethane). Diethyl ether is usually known simply as *ether*. This is the ether which was once widely used as an anaesthetic and which is today used as a solvent for substances which do not dissolve in water. It is very flammable and air/ether mixtures are dangerously explosive.

It is produced by the action of concentrated sulphuric acid on **ethanol**.

Ethylene See **Ethene**.

Ethyl ethanoate This is the **ester** produced by

reacting **ethanol** and **ethanoic acid**. It is used as a solvent in glues and paints.

Ethyne (or **acetylene**) A gaseous **alkyne** which is

$$H-C\equiv C-H$$

flammable and which forms explosive mixtures with **air**. It can be produced by the **cracking** of petroleum products, although it used to be made by the action of water on calcium carbide:

$$CaC_2(s) + 2H_2O(l) \rightarrow C_2H_2(g) + Ca(OH)_2(s)$$

The gas was once widely used for illumination, being burnt in special lamps. They were used in mines and on bicycles. Today ethyne is used to make **chloroethene** and thence **poly(chloroethene)** It is also used to produce very high **temperatures** by being burnt in the oxyacetylene torch used for welding and cutting metal. Temperatures of 2700°C can be obtained.

Evaporation This is a change of state from **liquid** to **vapour** which can occur at any temperature up to the boiling point. It takes place because molecules escape from the body of the liquid into the **atmosphere**. Only a small proportion of the molecules have sufficient energy to escape at any time but over a period they will all escape. Generally speaking, the *lower* the boiling point, the faster will be the rate of evaporation.

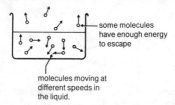

some molecules have enough energy to escape

molecules moving at different speeds in the liquid.

Exothermic reaction In an exothermic reaction heat energy is released to the surroundings from the reactants. The **bonds** of the products contain less energy than the bonds of the reactants. The products are *more* stable than the reactants.

Many common reactions are exothermic. All **combustion** and **neutralization** reactions are exothermic:

The **Haber process**

$$3H_2(g) + N_2(g) \rightarrow 2NH_3(g)$$

and **Contact process**

$$2SO_2(g) + O_2(g) \rightarrow 2SO_3(g)$$

are both exothermic reactions. **Enthalpy** changes (ΔH) are less than zero (negative). See **Endothermic reactions** and **Energy changes**.

92

Explosion A rapid expansion of **gas** which causes a shock wave. Common explosives are gunpowder, nitroglycerine and TNT. These materials react with oxygen if *detonated* and large temperatures are produced. A lot of gas is also produced and the gas molecules and the shock wave produced by the reaction travel outwards from the explosion with great energy. It is this energy which causes the damage in explosions.

In nuclear explosions the vast amount of energy locked within the **nucleus** of atoms is released.

Faraday (F) A Faraday is the amount of electric charge possessed by a **mole** of **electrons** (about 96500 **coulombs**). This amount of electricity is needed to liberate one mole of *atoms* of a *univalent element* (e.g. silver) during electrolysis:

$$Ag^+ \quad + \quad e^- \quad \rightarrow \quad Ag$$

One mole of silver ions.   One mole of electrons.   One mole of silver atoms.

Fehling's test This is a test which is used to detect certain **organic reducing agents**. Aldehydes can be distinguished from ketones, and sugars such as **glucose** from **starch** by using the test. Fehling's solution is a mixture of copper(II) sulphate, sodium hydroxide and sodium potassium tartrate. When it is heated with an appropriate reducing agent, e.g. glucose or ethanal, the copper (II) salt is reduced to a red precipitate of copper(I) oxide (Cu_2O).

Fermentation A process whereby **chemical changes** are made to **organic** chemicals by the use of living organisms such as **yeasts** and bacteria. The changes are brought about by **enzymes** acting as **catalysts**.

One important example is the changing of **sugars** to **ethanol** by the action of yeast:

$$C_6H_{12}O_6 \xrightarrow{\text{zymase}} 2C_2H_5OH + 2CO_2$$

glucose → ethanol + carbon dioxide

This reaction is used to make alcoholic drinks:

malted barley ⟹ beer
whisky (after **fractional distillation**)

fruit, e.g. grapes ⟹ wine
sherry and port
brandy (after fractional distillation)

Fertilizer A substance which is added to the soil to replace the nutrients which have been removed by plants. Fertilizers may be natural, e.g. manure and compost, or artificial, e.g. ammonium nitrate, superphosphate. Compounds containing **nitrogen, phosphorus** and **potassium** (N,P,K) are the main constituents of artificial fertilizers. The addition of fertilizers to the land is an important part of the **nitrogen cycle**.

Filter An insoluble **solid** and a **liquid** can be

separated by pouring the mixture into a filter paper in a filter funnel. The liquid (or solution) passes through the paper. This is known as the *filtrate*. The solid (known as the *residue*) is left in the filter paper. The whole process is known as *filtration*.

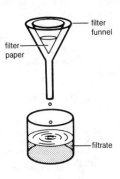

Fission This means *breaking up*. If a molecule is split into two or more parts it is said to undergo fission. In **nuclear reactions** the fission of a large atom, e.g. **uranium** into two smaller ones, e.g. **barium** and **krypton** releases enormous amounts of **energy**.

Fixing nitrogen Any process which converts atmospheric **nitrogen** into compounds which are useful as fertilizers. Fixing occurs in the **Haber**

process and is carried out by bacteria found on the roots of certain plants, e.g. peas and beans.

Flame When substances react and hot **gases** are produced the gases often give off light and **heat energy**. This is known as a flame. Flames vary in their colour and their temperature. The gas burner flame when the air hole is *closed* is yellow because the flame produced contains tiny glowing particles of unburnt carbon. The flame when the air hole is *open* (shown in the diagram) has different parts to it. The temperature varies in the different parts of the flame. An X marks the hottest place.

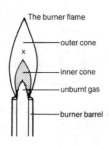

The burner flame
outer cone
inner cone
unburnt gas
burner barrel

Flame test This test can be used to identify metal ions present in compounds.

A clean nichrome or **platinum** wire is dipped into concentrated **hydrochloric acid** and then into a sample of the compound under test. The wire which now has a small amount of the compound attached to it is then put into the hottest part of the flame. Metals are characterized by specific colours:

Copper: green/blue   Potassium: lilac
Calcium: brick red   Barium: apple green
Sodium: yellow/orange   Lithium: bright red

Flask Containers (usually glass) in which chemical reactions take place. Some have graduations on them to show specific **volumes**.

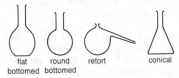

flat   round   retort   conical
bottomed   bottomed

Fluorine (F₂) A gaseous, nonmetallic element in group(VII) of the periodic table. It is a **halogen** and the most reactive element known.

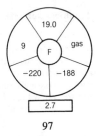

Fluorine is a vigorous oxidizing agent which will even oxidize chloride ion and water:

$$2Cl^-(aq) + F_2(g) \rightarrow 2F^-(aq) + Cl_2(aq)$$
$$2H_2O(l) + F_2(g) \rightarrow 4HF(g) + O_2(g)$$

Compounds of carbon and fluorine (fluorocarbons) are important refrigerants and aerosol propellants.

Fluorine is extracted by the **electrolysis** of a molten fluoride, e.g. KHF_2 and fluoride compounds are added to toothpaste to reduce tooth decay. See **Poly(tetrafluoroethene)**.

Formaldehyde (or **methanal)** A toxic gas belonging to the family of organic compounds called aldehydes. It is used extensively to make polymers and plastics like Bakelite, melamine and formica.

Formalin This is an **aqueous solution** of formaldehyde. It is used to preserve biological specimens.

Formula See **Empirical formula** and **Molecular formula**.

Fractional distillation (or **fractionation**) A **solvent** can be separated from a **solution** by the process of simple **distillation**. For a mixture of liquids with different **boiling points** it is usual to distil it and collect the products (*distillates*) which boil in definite temperature ranges. These distillates are termed *fractions.* The diagram on page 100 shows how **petroleum** might be fractionally distilled.

The exact temperature range of these fractions can be varied depending on the product which is in greatest demand. The distillation of whisky and brandy also involves fractional distillation.

Fructose A sugar molecule. Its formula is $C_6H_{12}O_6$ and it is found in Golden Syrup and honey.

Fuel A fuel is a substance which releases **heat energy** when it is treated in a certain way. In most fuels, the energy is released by **combustion**, e.g.:

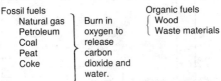

Fossil fuels
    Natural gas
    Petroleum
    Coal
    Peat
    Coke
Burn in oxygen to release carbon dioxide and water.

Organic fuels
    Wood
    Waste materials

Nuclear fuels, e.g. **uranium** and **plutonium**, produce heat because of changes that occur inside the atom. The **nuclear reactions** generate very large amounts of energy.

Functional group Refers to organic chemistry. It is the atom or group of atoms present in a molecule

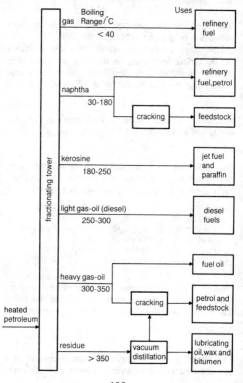

100

which is responsible for the characteristic properties
of that molecule.

e.g. —CO_2H carboxylic acid
      —CHO   aldehyde
      —NH_2 amine

The 'functional groups' of organic chemistry
correspond to the 'radicals' of inorganic chemistry.
See **Radical**.

Fusion This means *coming together* and has two
particular uses in chemistry.
  1) Fusion is another term for melting. See **Latent
heat**.
  2) Some **nuclear reactions** involve two atoms
coming together to form a single atom. This is the
opposite of **fission** and can involve the release of
enormous quantities of energy. Hydrogen bombs
are fusion weapons. Attempts are being made to use
fusion reactions to generate electricity.

(g) This is a **state symbol** and is used to denote that a
substance is a **gas**:

$CO_2(g)$, $O_2(g)$, $NH_3(g)$, $H_2O(g)$, $H_2(g)$

Galvanizing Iron and steel sheeting is often coated
with a thin layer of zinc. This is called *galvanized iron*.
Zinc is more resistant to corrosion than iron and so it
can protect the metal. Also, if the coating is scratched
and the iron and zinc come into contact with a
liquid and an electrochemical cell is set up, the zinc

101

reacts rather than the iron — it is more reactive. So, galvanizing protects the iron even when the protective layer is broken.

Gamma rays A form of high energy electro-magnetic radiation. They are produced in **nuclear reactions** and have great penetrating powers. They are similar to **X-rays**. They are used in the treatment of cancer — they kill body cells. They are also used to sterilize substances.

Gas A **state of matter** in which atoms and molecules have few bonds between them and consequently have a large amount of freedom of movement. They move at high velocity and in random directions.

When heat energy is supplied to a liquid, the atoms or molecules are given increased **kinetic energy**; this may be sufficient to overcome the bonds which hold them together in the **liquid** state. If this happens the liquid boils and turns into a gas or vapour. In a gas, atoms or molecules fill the whole container and their collisions with the walls of the container exert a **pressure**.

Gas laws If **Boyle's law** and **Charles' law** are combined we realize that for a *fixed mass of gas:*

$$\frac{(Pressure) \times (Volume)}{(Temperature)} = \text{constant}$$

$$\text{or:} \quad \frac{PV}{T} = \text{constant (T is in Kelvin)}$$

This means that if we compare gases under different conditions, e.g. different **temperature** and **pressure** then:

$$\frac{P_1 V_1}{T^1} = \frac{P_2 V_2}{T_2} = \text{constant}$$

condition 1   condition 2

This allows us to obtain much useful information about gases by calculation rather than by measurement. See **STP**.

General formula Refers to organic chemistry. This is a formula which shows the relative numbers of the different atoms in terms of the variable 'n' for all the members of a particular family of compounds. The actual formula of a particular compound is found by substituting for n. E.g. general formula of the alkanes is $C_n H_{2n+2}$.

Methane has 1 Carbon $\therefore$ n = 1 $\therefore$ formula = CH_4
Butane has 4 Carbons $\therefore$ n = 4 $\therefore$ formula = $C_4 H_{10}$

Giant structure **Molecules** are made up of small numbers of atoms bonded together. In giant structures of **atoms** or **ions** there are large numbers of particles present in a **crystal lattice**. Each particle has a strong force of attraction for all the other particles which are near to it. In this way attractive forces are spread through the structure and giant structures tend to have high **melting** and **boiling points**. Ionic substances have giant structures as do most elements, e.g. all metals and several nonmetals.

Glass A hard transparent mixture of silicates. The cheapest and commonest kind of glass is called *soda glass*. It is made by heating together sand, sodium carbonate, calcium oxide and broken glass (cullet). The hot liquid is cooled down very slowly. The slowness of cooling means that the glass does not crystallize. Glass is a *supercooled liquid*.

In *lead glass*, the sodium carbonate is replaced by lead(II) oxide. This gives a glass with a high refractive index. It is used in making crystal glassware. See **Borosilicate glass**.

Glucose A **monosaccharide** molecule. It is found in honey, Golden Syrup and fruits. All **sugar** and **starch** which enters our bodies is converted to glucose. It is then used to provide **energy**.

Glycol This material consists of ethane-1, 2-diol and is used as **antifreeze** material in engines. The compound contains two alcohol (OH) groups and is made from **ethene** by oxidation to epoxyethane and then water is added to give the diol. Ethene is obtained by the **cracking** of **petroleum**.

ethene   epoxyethane   ethane-1, 2-diol

Gold A valuable metal which is prized for its use as jewellery and often used as a substitute for money. It is found uncombined with other elements and the major deposits are in South Africa and the USSR.

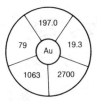

Chemically it is very *inert* reacting only with vigorous **oxidizing agents** such as **chlorine** and certain acids, e.g. *aqua regia*. Gold is a soft metal and for most uses it is alloyed with copper or silver.

Pure gold is said to be 24-carat gold. 9-carat gold is 9 parts gold to 15 parts copper, i.e. 37.5% gold. This is hard and is commonly used in jewellery.

Gram (g) A unit of **mass**. It is $1/1000$ of a kilogram and is used in all scientific work. The **symbol** (g) is used, e.g. 100 g.

Graphite This is an allotropic form of **carbon**. It is found naturally as *plumbago*. **Charcoal** consists of small particles of graphite. It is a **giant structure** with the carbon atoms bonded together in a hexagonal arrangement in layers or planes.

There are strong **bonds** between the atoms in the planes but the bonds between the atoms in different planes are weak. It is possible for the planes to slip over each other.

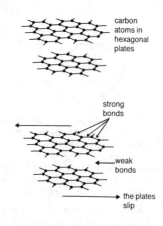

carbon atoms in hexagonal plates

strong bonds

weak bonds

the plates slip

Because of this property, graphite is very useful as a lubricant and, of course, in pencils where layers of graphite are left behind on the paper.

Graphite also finds use as electrodes in the extraction of elements, e.g. sodium, aluminium and chlorine, since it is an electrical conductor.

Group In the **periodic table** the **elements** are arranged in horizontal **periods** and vertical *groups*. The elements in these groups have similar chemical properties, e.g.

Group I	lithium, sodium, potassium
Group II	beryllium, magnesium, calcium
Group III	boron, aluminium
Group IV	carbon, silicon
Group V	nitrogen, phosphorus
Group VI	oxygen, sulphur
Group VII	fluorine, chlorine, bromine
Group O	helium, neon, argon

Each group has a characteristic **electronic configuration**, i.e. the outermost electron shell contains the same number of electrons for each member of the group. Furthermore this is the same as the group number.

Haber process **Ammonia** is made in this process from **hydrogen** and **nitrogen**. Nitrogen is obtained from the **air** and hydrogen from the steam **reforming** of **natural gas**. The gases are reacted together in a 1:3 ratio at 500°C and two hundred **atmospheres** **pressure**. An **iron catalyst** is used:

$$N_2(g) + 3H_2(g) \rightleftharpoons 2NH_3(g)$$

It is an **equilibrium reaction** and the conditions are chosen to produce as much ammonia as possible in the shortest time. Under these conditions about 15% of the reactants are converted to ammonia. Using a *lower* temperature will generate *more* ammonia but at a much slower **rate**. Conversely, using a *higher* temperature will produce the ammonia more quickly, but the **yield** will be lower. The chosen conditions are the *optimum* ones.

Haemoglobin The red pigment which is found in red blood corpuscles. It contains an atom of iron in the complex molecule. **Oxygen** reacts with the pigment forming *oxyhaemoglobin* and oxygen is released from the complex where it is needed. **Carbon monoxide** also reacts with haemoglobin, forming *carboxyhaemoglobin*. This is a very stable compound and prevents the carriage of oxygen. This is why carbon monoxide is such a dangerous poison. It is important that the body obtains sufficient iron each day to maintain the correct levels of haemoglobin. The daily requirement is about 11 mg. Insufficiency leads to the condition known as *anaemia*.

Half-life When a radioactive **isotope** gives off **alpha** or **beta particles** (decays) it changes into a different isotope. As the decay occurs the number of nuclei becomes fewer. The *half-life* of an isotope is the time taken for the **radioactivity** to decrease to half of its original value. The decay of an isotope is usually traced by measuring the **rate** at which particles are emitted. This rate is proportional to the number of nuclei present.

The plot shows the decay of an isotope with a half-life of one minute, i.e. in each minute the number

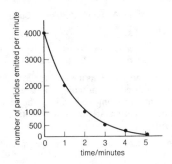

of particles emitted falls by half e.g. 4000–2000, 2000–1000, etc.

There is great variety in the length of half-lives from isotope to isotope. The half-life of each isotope is constant, however, under all conditions of **temperature** and **pressure**.

isotope	half-life
Carbon-14	5730 years
Oxygen-20	14 seconds
Copper-64	756 minutes
Uranium-234	250,000 years

Half-reaction It is often useful to represent a reaction by two half-reactions, e.g. the **displacement of copper ions** from solution by zinc:

$$Cu^{2+}(aq) + Zn(s) \rightarrow Zn^{2+}(aq) + Cu(s)$$

can be viewed as

(a) $Cu^{2+}(aq) + 2e^- \rightarrow Cu(s)$
(b) $Zn(s) \rightarrow Zn^{2+}(aq) + 2e^-$

From these half-reactions it is easy to see that the zinc is being oxidized by the copper ions and that the copper ions are being reduced by the zinc. Half-reactions are particularly useful for looking at **redox** reactions.

Halide A compound of a **halogen** with another **element**.
Examples:

hydrogen chloride (HCl) sodium bromide (NaBr)
calcium fluoride (CaF_2) silicon tetrachloride ($SiCl_4$)

Metal halides tend to be **ionic**, whereas nonmetal halides have **covalent bonds**. Metal halides can be produced by reacting the elements together:

$$2Na(s) + Cl_2(g) \rightarrow 2NaCl(s)$$

Halogens Elements in **group** VII of the **periodic table**. They are all poisonous, nonmetallic elements. **Fluorine** and **chlorine** are gases, **bromine** is a liquid and **iodine** is a solid at room temperature. They are all **oxidizing agents**. Their oxidizing power and chemical reactivity decreases in the order:

$$F_2 > Cl_2 > Br_2 > I_2,$$ i.e. as the group is descended.

They react vigorously with **metals** and **hydrogen**

forming **halides**. They all contain seven **electrons** in the outer **shell** of the atom and form univalent **anions**, e.g. Cl⁻ Br⁻.

Hardness of water Water from the mains supply is described as being *soft* or *hard.* Hard water does not easily lather with soap but forms a scum. It requires more soap than soft water which lathers easily. Hard water has passed over and through rocks which have dissolved in the water, e.g. **limestone, chalk** and gypsum ($CaSO_4$). Soft water has collected in areas of the country where the rocks are insoluble in water, e.g. the granite areas such as the Lake District, Cornwall and the Highlands of Scotland.

Hardness is caused by dissolved calcium and magnesium salts. The metal ions are actually responsible since they chemically react with the soap forming 'scum'.

There are two types of hard water:

Temporary: Contains dissolved calcium hydrogen-carbonate due to slightly acidic rain water dissolving chalk and limestone.

$$CaCO_3(s) + H_2O(l) + CO_2(g) \rightleftharpoons Ca(HCO_3)_2(aq)$$

| calcium carbonate | rainwater | calcium hydrogen carbonate |

It is called temporary since it is simply removed by boiling, which reverses the reaction to form the insoluble carbonate.

Permanent: Contains dissolved calcium sulphate and magnesium sulphate. These are *not* removed by boiling. They can be removed by adding sodium carbonate to precipitate the insoluble carbonates.

Soap forms a scum with hard water because of the formation of insoluble calcium salts. Soapless **detergents** do not do this. The use of kettles and immersion heaters with hard water becomes increasingly difficult and expensive because **calcium carbonate** is deposited on the heating elements making them inefficient. Substances such as *calgon* and *permutit* are used to make water softer through the use of **ion exchange**.

Heat energy   All substances possess heat energy. The higher their **temperature**, the more energy they possess. The source of this energy is the movement of atoms and molecules, i.e. their **kinetic energy**.

Heat energy is taken in during **endothermic reactions** and given out during **exothermic reactions**.

Helium (He)

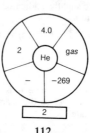

Helium is an **inert gas**. It is found in **natural gas** and is present in the **atmosphere** to a very small extent. It is completely unreactive and the gas is **monatomic**. It is used in airships as it is eight times less dense than air and nonflammable. Helium is also used in breathing apparatus for deep-sea divers.

Heterogeneous reaction   When chemical reactions occur between substances which are in different physical states, i.e. gases, liquids and/or solids, the reactions are said to be heterogeneous. Hetero ≡ different. See **Homogeneous reaction**.

Examples:

$$Zn(s) + 2HCl(aq) \rightarrow ZnCl_2(aq) + H_2(g)$$
$$2H_2O(l) + 2Na(s) \rightarrow 2NaOH(aq) + H_2(g)$$
$$2Na(s) + Cl_2(g) \rightarrow 2NaCl(s)$$
$$CuCO_3(s) \rightarrow CuO(s) + CO_2(g)$$

Hofmann voltameter   The voltameter is filled by opening the two taps and pouring the **electrolyte** into the central tube until the apparatus is full. When the taps are closed and the **current** switched on any gases which are produced will collect in the outside tubes above the electrodes.

The **volume** of the gases can be measured and the gases collected and tested.

See diagram over page.

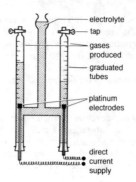

electrolyte

tap

gases produced

graduated tubes

platinum electrodes

direct current supply

Homogeneous reaction Reactions are said to be homogeneous when all the reactants and products are in the same physical state. Homo ≡ same.

Examples:

$$Fe(s) + S(s) \rightarrow FeS(s)$$
$$N_2(g) + 3H_2(g) \rightarrow 2NH_3(g)$$

Homologous series A series of compounds with the same **general formula**, e.g.

alkanes C_nH_{2n+2}
alkenes C_nH_{2n}

The compounds have the same **functional groups**

and have the same chemical **properties**. The physical properties of the series of compounds change only gradually.

Examples:
Alkanes: C_nH_{2n+2}

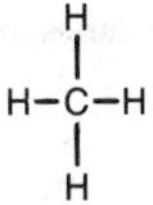

methane CH_4
n = 1

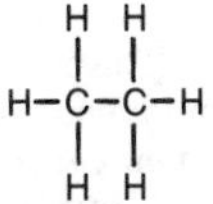

ethane C_2H_6
n = 2

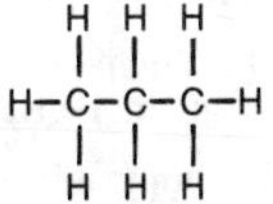

propane C_3H_8
n = 3

Hydrate A hydrate is a compound which has **water** chemically combined within it. *Hydrated* salts, i.e. those which contain **water of crystallization** are the best examples.

$$CuSO_4.5H_2O \qquad CaCl_2.6H_2O$$

Hydride A hydride is a compound which contains **hydrogen** and another **element** only. Examples:

NH_3 ammonia H_2O water H_2S hydrogen sulphide

The hydrides of nonmetals are **covalent** compounds. Some metal hydrides contain the **ion** H$^-$, e.g. Na$^+$H$^-$. These compounds are very reactive and will readily decompose water.

Hydrocarbon A compound of **hydrogen** and **carbon** only is a hydrocarbon, e.g. alkanes, alkenes, alkynes:

methane	CH$_4$	ethyne	C$_2$H$_2$
ethene	C$_2$H$_4$	benzene	C$_6$H$_6$

Hydrochloric acid (HCl(aq)) This is a solution of **hydrogen chloride** in **water** and it contains **chloride** and **oxonium** ions. The maximum **concentration** of the solution is 36% (about 11 **mol/dm^3**). The acid is **monobasic** and produces **salts** called **chlorides**, e.g.

$$Fe(s) + 2HCl(aq) \rightarrow FeCl_2(aq) + H_2(g)$$
$$Mg(s) + 2HCl(aq) \rightarrow MgCl_2(aq) + H_2(g)$$

It is a **strong acid** being fully dissociated into Cl$^-$(aq) and H$_3$O$^+$(aq) ions in dilute solution. It releases **carbon dioxide** from **carbonates** and **hydrogencarbonates** and can be oxidized to **chlorine**:

$$MnO_2(s) + 4HCl(aq) \rightarrow$$
$$MnCl_2(aq) + 2H_2O(l) + Cl_2(g)$$

Hydrogen (H$_2$) Hydrogen is a gaseous **diatomic element**. The atom consists of one **proton** and one **electron**. The **isotope deuterium** contains a **neutron** in the nucleus of the atom. There is a further isotope called Tritium which has 2 neutrons in its nucleus and is radioactive.

Hydrogen is very reactive. It can form **covalent bonds** by sharing electrons, e.g.

$$2H_2(g) + O_2(g) \rightarrow 2H_2O(g)$$
$$C_2H_4(g) + H_2(g) \rightarrow C_2H_6(g) \ (\textit{hydrogenation})$$

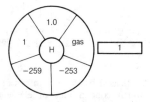

Two hydrogen ions can be produced by losing the electron to form H^+ or by gaining an electron to form H^- (present in some **hydrides**). The positive hydrogen ion (H^+) is such a small, reactive species (it is a lone proton) that it does not exist in solution. In **aqueous solution** it reacts with the water to form the **oxonium ion. Acids** contain the oxonium ion.

Hydrogen is a **reducing agent**. It is used to convert vegetable oils into *margarine*. Large quantities are used in the **Haber process**. Industrially, the gas is made from **petroleum** by the steam reformation of **alkanes**.

In the laboratory, the gas is made by reacting a metal with an acid other than nitric acid, e.g.:

$$Zn(s) + 2HCl(aq) \rightarrow ZnCl_2(aq) + H_2(g)$$

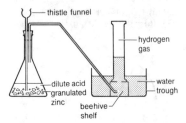

It is also released by the electrolysis of aqueous solutions which contain ions of elements above hydrogen in the electrochemical series, e.g. NaCl(aq), $Mg(NO_3)_2(aq)$, and by the reaction of water with alkali and alkaline earth metals, e.g. sodium, calcium.

Hydrogen is a flammable gas which causes explosive mixtures with oxygen. Great care must be taken in its preparation, collection and use.

Hydrogen bond A weak chemical bond can be formed between **hydrogen** atoms and atoms of **oxygen** or **nitrogen**. Although weaker than **covalent** or **ionic bonds**, hydrogen bonds affect the physical properties of compounds. Molecules in **water** and **ice** are extensively hydrogen-bonded to each other. See diagram on next page.

118

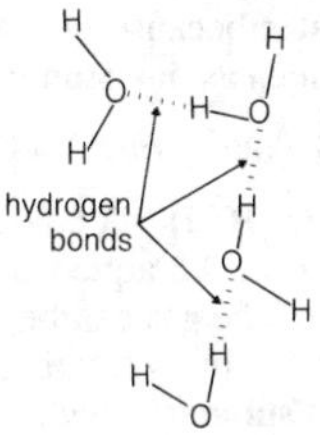

hydrogen bonds

Hydrogenation In this process an **unsaturated compound** is turned into a **saturated** one by the addition of **hydrogen**. A **catalyst** such as nickel is used, e.g.

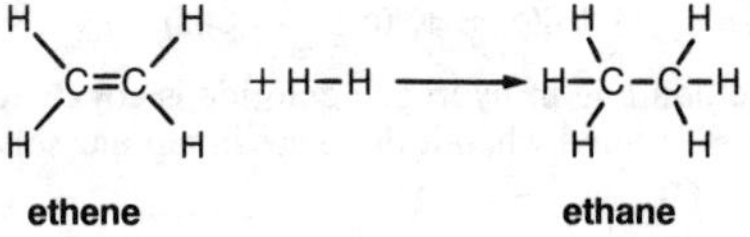

ethene + H–H ⟶ ethane

Hydrogencarbonates (HCO_3^-) These are the **acid salts** of *carbonic acid* H_2CO_3. The best known examples are:

Sodium hydrogencarbonate $NaHCO_3$. This is the familiar household chemical *bicarbonate of soda.* It is found in self-raising flour and **baking powder.**

Calcium hydrogencarbonate $Ca(HCO_3)_2$. This is

119

found in hard water because rain water passes over **carbonate** rocks such as **limestone**:

$$CaCO_3(s) + CO_2(aq) + H_2O(l) \rightarrow Ca(HCO_3)_2(aq)$$

Hydrogen chloride (HCl) A colourless gas which is very soluble in water. An aqueous solution is called **hydrochloric acid**. The gas can be made by reacting the elements together or by treating **sodium chloride** with concentrated **sulphuric acid**:

$$H_2(g) + Cl_2(g) \rightarrow 2HCl(g)$$
$$NaCl(s) + H_2SO_4(l) \rightarrow NaHSO_4(s) + HCl(g)$$

The gas will react with ammonia to form dense white fumes of **ammonium chloride**:

$$NH_3(g) + HCl(g) \rightarrow NH_4Cl(s)$$

The bonding in hydrogen chloride is **covalent** but **ions** are formed when it dissolves in a **polar** solvent, e.g. water.

Hydrogen halides These are gaseous compounds formed between **hydrogen** and the **halogens**. They are **covalent** compounds which readily dissolve in water to form an acidic solution which contains the **halide ion**, e.g.

$$HBr(g) + H_2O(l) \rightarrow H_3O^+(aq) + Br^-(aq)$$

Hydrogen fluoride	HF
Hydrogen chloride	HCl
Hydrogen bromide	HBr
Hydrogen iodide	HI

Hydrogen ion   The positively charged **hydrogen ion** is a **proton**:

$$H \rightarrow H^+ + e^-$$

This is such a small particle that it is very reactive. In solution, it is chemically combined with the solvent molecules. In water this is represented by the **oxonium ion**:

$$H^+ + H_2O \rightarrow H_3O^+$$

$$\left[\begin{array}{c} H \quad \quad H \\ O \\ | \\ H \end{array} \right]^+ \quad \text{oxonium ion}$$

The concentration of hydrogen ions in an aqueous solution is in **pH** units, and gives a measure of the degree of acidity of the solution. See **Hydroxide ion**.

Hydrogen peroxide (H_2O_2)   This compound is usually used as an **aqueous** solution. It readily decomposes to give oxygen:

$$2H_2O_2(aq) \rightarrow 2H_2O(l) + O_2(g)$$

A manganese(IV) oxide catalyst speeds up the reaction. It is used as a **disinfectant** and **bleach** in the home and is a powerful **oxidizing agent**. It bleaches hair to a blonde colour, and is used in industry for bleaching paper pulp and natural fibres. It is less destructive than the more powerful bleach chlorine.

Hydrogen sulphide (H_2S)   This is a colourless gas with a sweetish, sickly odour which reminds you of

121

rotten eggs. The gas is produced when **organic** matter containing **sulphur** rots. It is often found associated with petroleum. It is made in the laboratory by the action of an acid on a metal sulphide, e.g.:

$$FeS(s) + 2HCl(aq) \rightarrow FeCl_2(aq) + H_2S(g)$$

The gas is *very* poisonous. Like hydrogen cyanide [HCN] it prevents the transmission of energy within the body. The gas can be detected by the fact that it turns a piece of filter paper soaked in a lead(II) salt to a black colour. This is because insoluble lead(II) sulphide is formed:

$$Pb^{2+}(aq) + H_2S(g) \rightarrow PbS(s) + 2H^+(aq)$$
$$\text{black}$$

A solution of hydrogen sulphide in water is a **weak acid** and the gas is a **reducing agent**.

Hydrolysis This is the term for the **decomposition** of a substance by the action of water. The water is also decomposed. **Esters** are hydrolysed:

| ethyl ethanoate | water | **ethanoic acid** | **ethanol** |

Hydrophilic This term means *water-loving*. It is used to describe parts of molecules which readily dissolve in water, e.g. the ionic end of a **detergent** molecule is hydrophilic. This *attraction* for water is why detergents work. The other end of the molecule is **hydrophobic** — *water-hating*.

The hydrophobic part of a molecule does not dissolve in water and is not attracted to water. In a detergent this end of the molecule bonds with grease or oils in the material being cleaned. See **Soap** and **Detergent**.

Hydrophobic   See **Hydrophilic**.

Hydroxide ion   (OH^-)   This ion is found in all **alkalis**, e.g. sodium hydroxide, and in alkaline solutions. It is present in *all* aqueous solutions because of the **dissociation** of water:

$$H_2O \rightleftharpoons H^+ + OH^-$$

Solutions with *more* hydroxide ions than hydrogen ions are described as alkaline and have a **pH** greater than 7. Group I hydroxides are soluble in water. Some other hydroxides are sparingly soluble, e.g. $Mg(OH)_2$ $Ca(OH)_2$. Insoluble hydroxides can be precipitated by the use of a soluble hydroxide, e.g.

$$Pb(NO_3)_2(aq) + 2NaOH(aq) \rightarrow$$
$$Pb(OH)_2(s) + 2NaNO_3(aq)$$

Hygroscopic   A hygroscopic substance can take in up to 70% of its own mass of water without dissolving or getting wet. Examples are copper(II) oxide, sodium chloride and **silica gel**. See **Deliquescence**.

Ice Ice is solid, crystalline water. Its **melting point** at a **pressure** of one **atmosphere** is 0°C. The structure of ice is shown here. The regular, crystalline structure, means that frozen water can take on the patterns that we see in snowflakes. Ice is less dense than water and it floats.

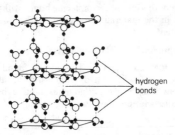

hydrogen bonds

Immiscible When two **liquids** do not mix together but form two layers — one liquid on top of the other, they are described as immiscible.

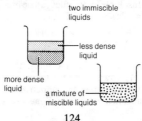

two immiscible liquids

less dense liquid

more dense liquid

a mixture of miscible liquids

One liquid will probably be polar, e.g. **water**, and the other non-polar, e.g. **ether**. Two polar liquids will mix completely, e.g. water and **ethanol**, as will two non-polar liquids, e.g. ether and **tetrachloromethane**. The two liquids may be separated using a separating funnel.

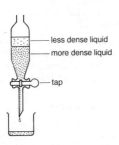

Indicator   An indicator is a substance which changes its colour under different conditions, e.g. litmus is red in acid solution and blue in alkaline. It can be used to *indicate* the **end point** of a reaction. Indicators are useful in **acid-base titrations** where they have different colours at different **pHs**. They are also used in titrations between **metal ions** and **complex ions**. See **Universal indicator**.

Inert gases See **Noble gases**.

Inorganic chemistry Inorganic chemistry is concerned with non-**organic** aspects of chemistry, i.e. **elements** and their **compounds**. This includes **carbon** the element, its **oxides**, metal **carbonates** and **hydrogencarbonates**. This excludes all organic compounds, e.g. **alcohols, esters, ethers, hydrocarbons**, etc.

Insoluble An insoluble substance is one which does not dissolve in the solvent.

Insulator An insulator is a substance which is a poor **conductor** of either heat or electricity.

Nonmetallic elements are usually insulators, as are most *solid* compounds and **polymers. Graphite** is an exception.

Examples of efficient insulators:

> Expanded polystyrene.
> Rubber.
> Mineral wools, e.g. *Rockwool*.
> Glass fibre.

Iodides Iodides are compounds of **iodine** and another **element**, e.g.:
Potassium iodide KI, Hydrogen iodide HI.

Iodine (I_2) A shiny, grey nonmetal. It is a **halogen** and has a **diatomic** molecular structure.

Iodine is extracted from sodium iodate(v) [$NaIO_3$] and a small amount from seaweed.

Silver iodide is used in photographic emulsions.

The human body needs 0.07 mg of iodine per day. Its principal use in the body is in the production of the hormone *thyroxine*. Potassium iodide is often added to table salt to provide this iodine.

Iodine will react with some metals directly to form **iodides**. It will react with **hydrogen** and **chlorine** forming [HI] and [ICl]. The vapour of the element is purple and is choking and **caustic**.

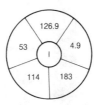

Ion A species which possesses an electrical charge. When an atom gains or loses an electron it becomes an ion. **Cations** have a positive charge, **anions** have a negative charge, e.g. sodium Na^+, oxide O^{2-}. Atoms tend to gain or lose **electrons** to produce an ion with an **inert-gas** configuration.

Groups of atoms (radicals) can also possess a charge, e.g.:

Sulphate SO_4^{2-}, Nitrate NO_3^-,
Hydroxide OH^-, Ammonium NH_4^+

They are also regarded as ions.

Ion-exchange It is possible to purify water by passing it over a resin (called an *Ion-exchange resin*) in a tube. With sea water the **sodium ions** would be replaced by **hydrogen** ions and the **chloride** ions by **hydroxide** ions. In other words sodium chloride would be *exchanged* for water.

The ions are bonded to the resin:

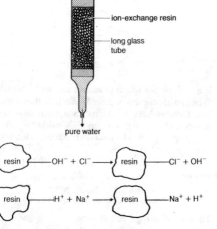

128

and exchange occurs. The product is *de-ionized* water. This method can be used for softening water, e.g. **calcium** ions are replaced by **sodium** ions.

Ionic bonds Chemical bonds which occur because of the electrostatic attractive forces between negatively and positively charged **ions**.

Ionic bonds occur in compounds of **nonmetals** from groups VI and VII and metallic elements, e.g.:

$$Na^+Cl^- \quad Mg^{2+}Br_2^-$$

and also in compounds involving radicals such as sulphate and nitrate, e.g.:

$$Cu^{2+}SO_4^{2-} \text{ and } K^+NO_3^-$$

Ionic compounds have **giant structures**.

Ionic equation The reaction between sodium hydroxide and hydrochloric acid can be described by the molecular **equation**:

$$NaOH(aq) + HCl(aq) \rightarrow NaCl(aq) + H_2O(l)$$

This reaction is an ionic one and can be represented in an *ionic* form:

$$Na^+OH^-(aq) + H^+Cl^-(aq) \rightarrow$$
$$Na^+Cl^-(aq) + H_2O(l)$$

Because Na^+ and Cl^- appears on *both* sides of the equation they do not take any part in the reaction and can be omitted. The *ionic* equation is therefore:

$$OH^-(aq) + H^+(aq) \rightarrow H_2O(l)$$

The other ions are **spectator ions**.

Ionization When an **atom** loses or gains **electrons** it becomes an **ion**. This gain or loss is termed 'ionization'.

Iron The most widely used metallic element. It is mainly encountered as **steel alloys** and in this form it is used for building girders, machine bodies (cars, cookers, fridges), containers (boxes, cans, drums), tools, utensils and many other items of everyday life.

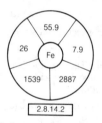

2.8.14.2

The metal is extracted from ores such as haematite (Fe_2O_3) and magnetite (Fe_3O_4) by reduction with **carbon monoxide** in the **blast furnace**. The ion which is produced in this way is brittle and is made into many different kinds of steel to strengthen it and to give it special properties.

130

One of the main problems with iron is that it rusts, i.e. it oxidizes in air to produce a soft, crumbly oxide.

Iron (strong, useful metal) $\xrightarrow{\text{(moist air)}}$ Iron(III) oxide (Fe_2O_3) (weak, worthless oxide)

Iron is a **transition metal** and can have **valency** 2 or 3. The metal reacts with dilute acids to form iron(II) compounds, e.g.:

$$Fe(s) + 2HCl(aq) \rightarrow FeCl_2(aq) + H_2(g)$$

but will give iron(III) compounds when reacted with vigorous **oxidizing agents** such as **chlorine**:

$$2Fe(s) + 3Cl_2(g) \rightarrow 2FeCl_3(s)$$

Iron has a **reversible reaction** with steam:

$$3Fe(s) + 4H_2O(g) \rightleftharpoons Fe_3O_4(s) + 4H_2(g)$$

Iron(III) compounds can be identified by their giving a red colour with potassium thiocyanate solution (KCNS). Iron(II) compounds do not do this but do produce a blue colour with a solution of potassium hexacyanoferrate(III) $K_3Fe(CN)_6$. Furthermore, when reacted with sodium hydroxide solution, Iron(II) solutions produce a muddy green precipitate, whereas Iron(III) solutions produce a rust-brown precipitate.

Iron compounds Iron forms compounds with

131

valency (II) and (III). Iron(II) compounds tend to be green in colour whilst iron(III) compounds are yellow or brown.

Iron(III) oxide Fe_2O_3 (haematite)	This is used as a yellowish/brown pigment in the paper, linoleum and ceramics industries. It is used as a **catalyst** and as a polish, e.g. in jeweller's rouge.
Iron(III) chloride $FeCl_3$	This is produced by reacting the elements together. It is an important catalyst and also finds use in the purification of water and in the production of pharmaceutical products.
Iron(II) sulphate $FeSO_4.7H_2O$	Iron(II) sulphate can be recovered from the waste materials left behind in the electroplating processes. It is an important compound, used in the preservation of wood, in inks and in lithography.

Isomer When two or more *different* compounds have the same **molecular formula** they are described as isomers.

Because the compounds are different, they have different properties.

Example:

Formula C_2H_6O

Structure of the compounds

ethanol   methoxymethane

Isotopes   Atoms of an **element** always contain the *same* number of **protons** (same **atomic number**). The number of **neutrons** in atoms of an element can be different (different **mass number**).

A sample of **chlorine** contains atoms which have eighteen neutrons (76% of the total) and atoms which have twenty neutrons (24%). Thus, chlorine is said to have two isotopes:

$$^{35}_{17}Cl \quad ^{37}_{17}Cl$$

Thus isotopes are atoms of the same element containing different numbers of neutrons.

The ratio of the isotopes is always constant. Most elements are found in *isotopic forms*. Some examples are found here:

	Isotopes	
Bromine	$^{79}_{35}Br$ (51%)	$^{81}_{35}Br$ (49%)
Carbon	$^{12}_{6}C$ (99%)	$^{13}_{6}C$ (1%)

Copper	$^{63}_{29}$Cu (69%)	$^{65}_{29}$Cu (31%)
Magnesium	$^{24}_{12}$Mg (79%)	$^{25}_{12}$Mg (10%)
	$^{26}_{12}$Mg (11%)	

All the ones quoted are *stable* isotopes. However, many more unstable, radioactive isotopes exist. A few elements have no isotopes, i.e. they have no variation in the number of neutrons that are found in the atom, e.g.:

Fluorine	Manganese
Gold	Phosphorus
Iodine	Scandium

The existence of isotopes accounts for the fact that many elements have a **relative atomic mass** which is not close to a whole number e.g. Cl = 35.5.

Joule The **SI unit** of **energy** and work. It has the symbol J. In scientific usage the **kilojoule** (kJ) is common.

Kelvin temperature scale The Kelvin is the **SI unit** of **temperature**. The Kelvin (K) is the same as one degree **Celsius**. Absolute zero is 0K.

Here is a comparison with the Celsius scale:

Kelvin	0	273	310	373	K
Celsius	−273	0	37	100	°C
	absolute zero	freezing point of water	blood heat	boiling point of water	

134

Note: A temperature in Kelvin does *not* include a degree sign (°), thus the boiling point of water is simply 373K.

Kerosine A product of **petroleum** refining. Its uses include fuel for jet aircraft and also household paraffin which can be used for heating and lighting. Kerosines boil between 160°C and 250°C.

Kilo A prefix which means *one thousand*. In common usage, a kilo has come to mean a kilogram, e.g.: 'a kilo of potatoes'. This is imprecise and should not be used in scientific work. Uses:

kilogram	=	1000 g
kilometre	=	1000 m
kilojoule	=	1000 J
kilowatt hour	=	1000 Wh

Kilojoule (kJ) One thousand **joules**.

Kinetic energy **Energy** a body possesses because of its motion. The greater its velocity (speed) the more energy it has. The kinetic energy of a particle is $\frac{1}{2} mv^2$ where m is its **mass** and v its velocity. If the mass is measured in kilograms and the velocity in metres/second the kinetic energy is measured in **joules**.

Kinetic theory The kinetic theory explains that all particles (**atoms** and **molecules**) are moving and that the extent to which movement can occur depends on the temperature. In **solids** and **liquids** the amount of movement is restricted by bonds between adjacent

135

particles. In **gases** the movement is only restricted by the walls of the container. The higher the temperature, the greater the **kinetic energy** the particles possess. See **Lattice**.

Krypton A **noble gas**. It is found in the **atmosphere**

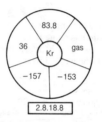

to a small extent and is used for electronic valves and fluorescent tubes. Like all noble gases, krypton is **monatomic**.

l 1. A **liquid**, e.g. $H_2O(l)$, $Hg(l)$, $H_2SO_4(l)$.
2. The unit of volume litre which is equivalent to 1 dm^3 or 1000 ml.

L This is the symbol for **Avogadro constant**. It is the number of particles in one **mole** of a substance. Its value is 6×10^{23} particles per mole.

Latent heat The amount of **heat energy** released or absorbed in a change of state at a fixed **temperature**

(e.g. **melting point** and **boiling point**). The latent heat of *fusion* is that energy needed to turn one **mole** of **solid** into a **liquid** *at its melting point*.

Similarly, the latent heat of *vaporization* is that energy required to turn one mole of **liquid** into a **gas** at its boiling point. These symbols can be used:

ΔH_m m = melting (fusion)

ΔH_b b = boiling (vaporization)

Examples:

Water	ΔH_m	=	6kJ/mole
	ΔH_b	=	41 kJ/mole
Magnesium	ΔH_m	=	9 kJ/mole
	ΔH_b	=	129 kJ/mole

Lattice A lattice is a regular arrangement of **molecules, atoms** or **ions** within a crystalline solid. Lattices contain large numbers of particles which are arranged in very particular ways. Small sections of three lattices are shown below. They are (a) aluminium (b) sodium (c) magnesium.

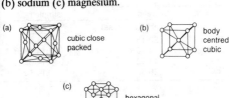

(a) cubic close packed

(b) body centred cubic

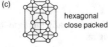

(c) hexagonal close packed

137

Law of constant composition A chemical **compound** always contains the same **elements** combined in the same proportions by **mass**.

Examples:

Water always has hydrogen and oxygen within it in the proportion: $^H/_O = ^2/_{16} = ^1/_8$ by mass

Carbon dioxide always has carbon and oxygen within it in the proportion: $^C/_O = ^{12}/_{32} = ^3/_8$ by mass

Law of multiple proportions If two **elements** form more than one **compound** then there is a simple relationship between the masses of one of the elements and a *fixed* mass of the other element.

Example: Oxides of nitrogen:

formula	ratio	simple ratio
	N:O (by mass)	N:O
N_2O	28:16	14:8
NO	14:16	14:16
NO_2	14:32	14:32
N_2O_5	28:80	14:40

There is a simple and steady increase in the ration of oxygen to nitrogen.

Lead A dense metallic **element** which is in group IV of the **periodic table.** It is a soft, **malleable**, grey element which is extracted from *galena* (PbS). Lead and its **alloys** are widely used, e.g. in the lead-acid **accumulator**, solders, protection from moisture (used

on roofs), radioactive isotopes and X-rays. Because of its chemical inertness, lead can be used to store **sulphuric acid**.

Lead compounds Lead forms lead(II) and lead(IV) compounds. Most lead compounds are cumulative **poisons**.

Lead(II) nitrate $Pb(NO_3)_2$	This is the only common **soluble** lead compound.
Lead(II) oxide PbO	This is the yellow oxide, litharge. It is used in the making of glass and enamels.

139

| Lead(IV) oxide PbO_2 | This oxide is formed in the lead-acid **accumulator**. |
| Tetraethyl lead $Pb(C_2H_5)_4$ | This is the 'antiknock' agent added to petrol to make car engines run smoother. It is also a source of air pollution from car exhausts. |

Le Chatelier's principle If a reaction is at **equilibrium** and any of the conditions are changed, further reaction will occur to counter the changes and re-establish equilibrium.

Examples:

The **Haber process**: $N_2(g) + 3H_2(g) \rightleftharpoons 2NH_3(g)$

If extra **nitrogen** or **hydrogen** is added to an equilibrium mixture then more **ammonia** will be formed and the equilibrium will be re-established.

Similarly, if extra ammonia is added, more nitrogen and hydrogen will be formed.

If the **pressure** is *increased* more ammonia will be formed. This is because forming more ammonia leads to a *lowering* of the pressure due to an overall reduction in the number of gas molecules present. In this way equilibrium is re-established.

A change of **temperature** leads to a change in the proportions of the equilibrium mixture. For **exothermic** reactions (e.g. Haber process) *raising* the temperature favours the *right → left* reaction (ammonia would react to form more nitrogen and

140

hydrogen). This is because the right → left reaction is **endothermic**. It therefore counters the rise in temperature by absorbing heat.

Lime (or quicklime) This is **calcium oxide** (CaO). It is produced by heating **calcium carbonate**. It is used to make **slaked lime** and is responsible for turning impurities in the **blast furnace** into *slag*. In the laboratory it is a useful drying agent for **ammonia**.

Limestone A commonly found rock which contains between 50% and 90% **calcium carbonate** ($CaCO_3$). It is used to make **lime** and **cement** and is used as building stone and hard core for foundations.

Limewater A dilute solution of the sparingly soluble compound **calcium hydroxide**. It is an **akali** and is used to test for **carbon dioxide**:

$$Ca(OH)_2(aq) + CO_2(g) \rightarrow CaCO_3(s) + H_2O(l)$$

A white precipitate is formed seen as a milkiness in the solution. If excess carbon dioxide is bubbled through, the precipitate reacts to form the soluble salt **calcium hydrogencarbonate** and the liquid goes clear:

$$CaCO_3(s) + H_2O(l) + CO_2(g) \rightarrow Ca(HCO_3)_2(aq)$$

Linear molecules Molecules which are *straight*, i.e. their **atoms** are *in a line*. All molecules which contain only two atoms must be linear but some examples of more complicated molecules are shown here:

carbon dioxide	ethyne	beryllium chloride
O = C = O	H — C ≡ C — H	Cl — Be — Cl

Liquid A **state of matter** in which particles are loosely bonded by intermolecular forces. A liquid always takes up the shape of its container. The particles in the liquid are not fixed in a rigid framework (**lattice**).

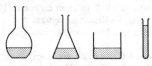

See **evaporation** and **boiling**.

Lithium A group I **metal**. It is the least reactive

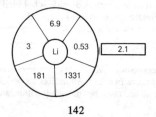

element in the group. Nevertheless it is stored under oil because of its reactivity towards air and water. It is soft, and can be cut with a knife, revealing a silvery surface which tarnishes readily. It has a steady reaction with water but reacts vigorously with acids.

$$2Li_{(s)} + 2H_2O_{(l)} \rightarrow 2LiOH_{(aq)} + H_{2(g)}$$

It is possible that the metal will be very important in the future if electricity can be generated by **nuclear fusion** reactions.

The metal ion gives a red **flame test**.

Litmus   This is extracted from lichen. It is used as an **acid-base indicator**:

acidic solution	pH 7	alkaline solution
red	purple	blue

Litre   A litre is exactly 1 **dm³** (1000 cm³). It is the **unit** of **volume** in everyday life, e.g. for fruit juice and paint, but the *scientific unit* is the cubic decimetre, **dm³**.

Lone pair of electrons   These are pairs of *outer shell electrons* which are not used in bonds within the compound. They can, however, be used for forming bonds with other compounds. See example over page.

Example:

Three bonding pairs of electrons

Here the lone pair is used to form the ammonium ion

M See **Molarity**.

M$_r$ This is the symbol for the **relative molecular mass** of a **compound** or **molecule**. It can be calculated by adding together the **relative atomic masses (A$_r$)** of each atom within the molecule.

Example:

M$_r$ carbon monoxide (CO) $= 12 + 16 = 28$
M$_r$ water (H$_2$O) $= (2 \times 1) + 16 = 18$

Macromolecule This term means *large* molecule. It is usually used to describe **molecules** with an **M$_r$** value greater than one thousand, i.e. **polymers** such as the **carbohydrates** and the plastics, e.g. **poly(thene)**, and also molecules such as **proteins** and nucleic acids.

Magnesium Magnesium is a shiny grey **group II metal**. It is quite reactive giving vigorous reactions towards **acids**. It burns vigorously in air with a bright white light, hence its use in flares, fireworks and flash bulbs. It also burns in carbon dioxide gas, and will

144

react with steam to release hydrogen. The metal is obtained by the **electrolysis** of molten magnesium chloride. Much of this is extracted from seawater.

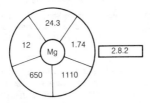

The chief use of magnesium is in the production of low density **alloys** for use in the aircraft industry.

Magnesium compounds

Magnesium hydroxide $Mg(OH)_2$	This is obtained by treating sea-water with calcium hydroxide. It is used in pharmaceutical products such as *Milk of Magnesia* as a treatment for excess acidity in the stomach.
Magnesium carbonate $MgCO_3$	This is found naturally as magnesite and as dolomite ($MgCO_3.CaCO_3$) and is used for making heat-resistant (refractory) materials.
Magnesium oxide MgO	This important refractory compound is made by heating the carbonate.

| Magnesium chloride $MgCl_2$ | After extraction from seawater the molten compound is electrolysed to produce magnesium metal. |
| Magnesium sulphate $MgSO_4.7H_2O$ | *Epsom Salts.* These crystals are used to treat constipation. They are also used as a fire-proofing agent. |

Malleable A malleable substance can be beaten or hammered into different shapes. The larger the crystals in the substance, the more malleable it is. Materials which are malleable are usually also **ductile**. **Metals** and **alloys** have these properties.

Maltose ($C_{12}H_{22}O_{11}$) This is a **disaccharide** molecule. It is found as a breakdown product of **starch**. Maltose is a **glucose dimer**.

Manganese Manganese is a **transition metal**. It is found naturally as the oxide (MnO_2) and is extracted either by electrolysis of the sulphate or by a **thermit reaction**. The chief use of the metal is in **alloys**, e.g. **steels** and **bronzes**.

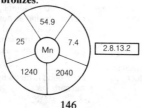

Manganese compounds Manganese forms compounds where the metal has valency 2, 3, 4, 6 and 7. The most important ones are 2, 4 and 7.

Manganese(IV) oxide MnO_2	This has an important use in the dry **battery**. It is used as a **catalyst** and oxidizing agent.
Potassium manganate(VII) $KMnO_4$	This is a vivid purple crystalline substance which is a vigorous **oxidizing agent**. It is used as an antiseptic. It gives the MnO_4^- ion.

Mass This is how much material a substance possesses. It is usually measured in **grams** (g) or kilograms (kg)

Mass number (A) The mass number of an **isotope** is the sum of the number of **protons** and **neutrons** in the **nucleus** of the atom. This number is shown at the top left-hand side of the symbol when describing the isotope (the number below it being the atomic number).

Examples:
$${}^{1}_{1}H \quad {}^{12}_{6}C \quad {}^{23}_{11}Na \quad {}^{40}_{20}Ca \quad {}^{208}_{82}Pb \quad {}^{238}_{92}U$$

Melting A **solid** melts to form a **liquid** when the **energy** of the particles is sufficient to break up the bonds holding them in a **lattice**. For a *pure* substance, melting occurs at a fixed temperature — the **melting point**. Some bonds remain but clusters of particles are mobile.

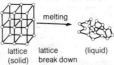

lattice (solid)   lattice break down   (liquid)

Melting point The temperature at which a solid melts to form a liquid (or a liquid solidifies to form a solid). More precisely, it is the temperature at which solid and liquid forms of the same substance (e.g. ice/water) are in equilibrium. At constant pressure, the melting point is a constant for a **pure** substance but it is *lowered* if impurities are added, hence ice melts when sprinkled with salt.

Mercury Mercury is the only liquid **metal** at room temperature. It is used to make **amalgams**, in **electrolysis** cells as a **cathode**, and in thermometers. Its compounds are very poisonous as is its vapour.

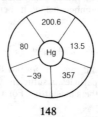

Metal 17% of elements are metals. A list of their general properties is given below. Pure metals are rarely used today. Most metals are used in the form of **alloys**. These play a vital part in our lives.

Metals produce **cations**, react with **acids**, are strong and hard, are **malleable** and **ductile**, are good conductors of heat and electricity and react with nonmetals. They have a crystalline structure, are shiny and have high **melting** and **boiling points**.

Metalloid These elements are neither **metals** nor **nonmetals**. They are sometimes referred to as *semimetals*.

The best examples are germanium and **arsenic**. Such substances tend to have the physical properties of metals, e.g. have a shiny appearance and a high melting and boiling point but the chemical properties of nonmetals, e.g. do not react with acids. They are useful as **semiconductors**.

Methane A gaseous **alkane**. It is the main constituent of **natural gas**. It is also released from **petroleum** whilst it is being processed. It burns readily to give carbon dioxide and water and is an industrial source of **hydrogen**.

CH_4

Methanol An **alcohol**. It is a colourless poisonous

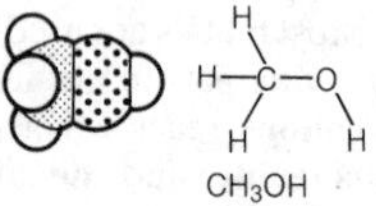

CH₃OH

liquid. It is added to **ethanol** to make **methylated spirits**.

Methylated spirits This is a mixture of **ethanol** (90%), **methanol** (9.5%), pyridine (0.5%) and traces of a purple dye. This mixing is done to make the ethanol unfit for drinking. There is no excise tax on the mixture and so it is inexpensive. It is used as a fuel and a **solvent**.

Methylbenzene (formerly **toluene**) This is an **aromatic hydrocarbon** which is produced from petroleum. It is used to make the explosive TNT (TriNitro Toluene) and is a useful **solvent**.

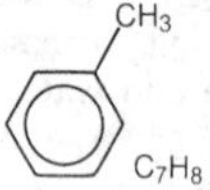

CH₃

C₇H₈

Methyl orange This is an **acid-base indicator**.

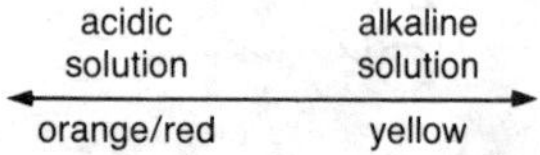

acidic solution	alkaline solution
orange/red	yellow

Miscible liquids These are liquids which can mix together completely. They can dissolve in each other, e.g. water and ethanol.

They require **fractional distillation** to separate them. See **Immiscible**.

Mixture When two or more substances are present in the same container they are said to be a mixture, e.g. **baking powder, air**, gunpowder, **petrol, methylated spirits**.

Properties of mixtures:
  Mixtures can be made in all proportions.
  Making mixtures does not involve the release or absorption of heat.
  The properties of a mixture are the properties of *all* the components.
  Mixtures can be separated by physical means.

ml Abbreviation for millilitre.

mmHg This symbol represents the **unit** *millimetres of mercury*. It is a unit of **pressure**. One **atmosphere** pressure equals 760 mmHg. This is the **air** pressure which would support a column of mercury 760 mm high.

Molarity (M) The molarity of a **solution** is its **concentration** expressed in **moles** per cubic decimetre of solution (**mol/dm³**). A solution containing 2 moles per dm³ is expressed as 2M. See **Molar solution**.

Molar solution A molar solution is one where the **concentration** of **solute** is *one* **mole** per cubic decimetre of *solution* 1 **mol/dm³**, i.e. a solution where one mole of solute has been dissolved in the **solvent** and then sufficient solvent has been added to make 1000 **cm³** of solution.

Mole This is the amount of a substance (**element** or **compound**) which contains **L** particles (L is the **Avogadro constant**).

L is defined as the number of atoms there are in 12 g of the **carbon-12 isotope**. It follows that the mass of one mole of an element or compound is the **relative atomic mass** (expressed in grams).

Examples:
One mole of

Oxygen	(O_2)	is $16 + 16$	$= 32$ g
Water	(H_2O)	is $2(1) + 16$	$= 18$ g
Ammonia	(NH_3)	is $14 + 3(1)$	$= 17$ g
Ethene	(C_2H_4)	is $2(12) + 4(1)$	$= 28$ g

mol/dm³ This is a **unit** of **concentration**. 1 mol/dm³ means that one **mole** of substance would be present in one cubic decimetre (**dm³**) of the **solution**. The unit is sometimes written mol dm⁻³ and often given the symbol M, e.g. 2 M. See **Molarity**.

Molecule A molecule is defined as the smallest particle of an **element** or **compound** which exists independently. It contains **atoms** bonded together in a fixed whole number ratio:

Oxygen O_2	Phosphorus P_4
Nitrogen N_2	Carbon dioxide CO_2
Neon Ne	Water H_2O
Sulphur S_8	Hydrogen chloride HCl
Hydrogen H_2	

All organic compounds are molecules.

Molecular formula The molecular formula of a substance shows the number and types of atoms in the molecule. It tells us nothing about how the atoms are arranged, e.g. $C_4H_{10}O$ is the molecular formula of butanol and **ether** (ethoxyethane). See **Empirical formula** and **Structural formula**.

Monatomic molecule A **molecule** which only contains one **atom**. Only the **noble gases** are monatomic. Because of their **electronic configurations** it is difficult for inert gas atoms to form bonds, and so they exist in the monatomic form.

Monobasic acid A monobasic acid only contains *one* **hydrogen** atom per molecule which can be replaced by a metal. Only normal salts can be formed — no **acid salts**.

Examples include:
Hydrochloric acid HCl
Nitric acid HNO_3

Ethanoic acid

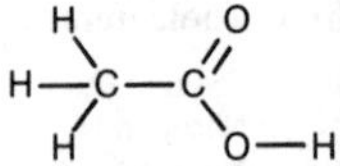

$$H-\overset{\displaystyle \overset{H}{|}}{\underset{\displaystyle \underset{H}{|}}{C}}-C\overset{\displaystyle O}{\underset{\displaystyle O-H}{}}$$

Monomer The compound from which **polymers** are made, e.g. **ethene** $CH_2 = CH_2$ forms **poly(ethene)** $(CH_2 - CH_2)_n$.

Monosaccharide The simplest kind of **sugar**. In these compounds there is a *single* molecule as opposed to two (or more) which have reacted together to form **disaccharides** or **polysaccharides**.

Glucose	$C_6H_{12}O_6$
Fructose	$C_6H_{12}O_6$
Ribose	$C_5H_{10}O_5$

Multiple bond **Bonds** which contain two **electrons** as a shared pair are called **single bonds**. Those which contain four or six electrons i.e. two or three shared pairs are multiple bonds. **Double bonds** contain four electrons. **Triple bonds** contain six.

Examples of compounds which contain multiple bonds:

Ethene C_2H_4 — **double bond**
Ethyne C_2H_2 — **triple bond**

154

Nanometre   This is a unit of length. It is 10^{-9} of a metre, i.e. 1 000 000 000 make one metre. It is useful in measuring the length of bonds, e.g. H—H bond length = 0.074 nm.

Naphtha   This is a mixture of **hydrocarbons** which is produced by the **fractional distillation** of **petroleum**. The **boiling points** of the compounds in the mixture are in the range 80—160°C. Naphtha is usually subjected to **cracking**. **Ethene** is an important product of this process. Naphtha can be oxidized to give **ethanoic acid**.

Natural gas   This is a mixture of mainly **hydrocarbon** gases which is found in deposits beneath the earth's surface. Natural gas and petroleum are often found together. **Methane** is usually the major constituent. The **inert gas, helium**, is sometimes present in the mixture.

It is used as a fuel both in industry and the home.

Neon   Neon is a **noble gas** which is found in the **atmosphere** to a small extent. It forms no known compounds but is used to fill fluorescent tubes. A red glow is produced which is used for advertising signs. See diagram over page.

155

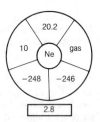

Neutral **1.** A neutral **solution** is one which is neither acidic nor alkaline. Neutral solutions contain the same concentration of **hydroxide** and **oxonium ions**. The pH of a neutral solution is seven at 25°C.

2. A neutral **oxide** does not react with either **acids** or **alkalis**. The best example is **water**.

3. A neutral particle is one which does not have an electric charge, e.g. a **neutron**.

Neutralization This is the process in which either the **pH** of an acidic solution is increased to seven or the pH of an alkaline solution is decreased to seven. The resulting solution contains the same concentration of **oxonium** and **hydroxide** ions i.e. it is neutral.

An acid can be neutralized by the addition of a base or a compound such as a carbonate.

$$HCl(aq) + NaOH(aq) \rightarrow NaCl(aq) + H_2O(l)$$
$$2HCl(aq) + CaCO_3(s) \rightarrow$$
$$CaCl_2(aq) + CO_2(g) + H_2O(l)$$

Acid/base **indicators** show when neutralization is complete.

Neutron This is one of the particles which are found in the **nucleus** of all atoms except hydrogen. It has approximately the same mass as the **proton** but *no* charge. See **Nuclear reactions**.

Nickel Nickel is a transition metal. It is a magnetic substance which is found in nature as the **sulphide** and is oxidized to the **oxide**, reduced by **hydrogen** and

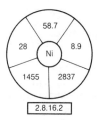

then purified by using carbon monoxide gas:

$$Ni(s) + 4CO(g) \rightarrow Ni(CO)_4(g)$$

This reaction can be reversed at high temperatures, producing **pure** nickel.

Nickel can have a **valency** of 2, 3 or 4, but only nickel(II) salts are common. Nickel is used as a **catalyst**, in **alloys** (nichrome, coinage metal, stainless steel) and in plating.

Nitrate Nitrates contain the $-NO_3$ group with **valency = 1**. These are **salts of nitric acid** and can be made by treating a metal **carbonate** or **oxide** with the dilute acid:

$$CuCO_3(s) + 2HNO_3(aq) \rightarrow$$
$$Cu(NO_3)_2(aq) + CO_2(g) + H_2O(l)$$
$$ZnO(s) + 2HNO_3(aq) \rightarrow$$
$$Zn(NO_3)_2(aq) + H_2O(l)$$

Nitrates are easily decomposed by heat. There are three kinds of reaction:

(a) $2KNO_3(s) \rightarrow 2KNO_2(s) + O_2(g)$
examples: *Sodium, Potassium.*
(b) $2Cu(NO_3)_2(s) \rightarrow 2CuO(s) + O_2(g) + 4NO_2(g)$
all metal nitrates other than shown in (a) and (c).
(c) $2AgNO_3(s) \rightarrow 2Ag(s) + 2NO_2(g) + O_2(g)$
examples: *Silver, Mercury.*

Nitrates are important fertilizers, e.g. **sodium nitrate** and **ammonium nitrate**. If used incorrectly they can lead to serious water pollution.

Nitric acid (HNO_3) This colourless, corrosive liquid is made from **ammonia** by **oxidation** over a **platinum/rhodium catalyst**. There are three stages:

$$4NH_3(g) + 5O_2(g) \rightarrow 4NO(g) + 6H_2O(g)$$
$$4NO(g) + 2O_2(g) \rightarrow 4NO_2(g)$$
$$4NO_2(g) + 2H_2O(l) + O_2(g) \rightarrow 4HNO_3(aq)$$

The acid is distilled to a concentration of 68% and can be prepared in the laboratory by heating a nitrate with *concentrated* **sulphuric acid**:

$$KNO_3(s) + H_2SO_4(l) \rightarrow KHSO_4(s) + HNO_3(g)$$

Nitric acid is a vigorous oxidizing agent used in the production of **fertilizers** and explosives. The **salts** of nitric acid are termed **nitrates**. See **Nitrogen oxides** and **Nitrites**.

Nitrites These are salts of nitrous acid (HNO_2). The sodium and potassium salts can be made by heating the nitrate:

$$2NaNO_3(s) \rightarrow 2NaNO_2(s) + O_2(g)$$

They are used in the preservation of meats.

Nitrogen (N_2) A nonmetallic element in group v of the periodic table. It is an unreactive, **diatomic** gas which forms about 78% of the **atmosphere**. It is produced by the **fractional distillation** of liquid **air**.

Nitrogen is an important chemical because of the need for **nitric acid** and **ammonia**. It is *fixed* from the atmosphere in the **Haber process**.

See **Nitrogen oxides** and **Nitrogen cycle**.

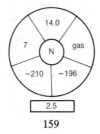

Nitrogen cycle The diagram opposite shows the ways in which animals, plants and humans are involved in the chemical links between **nitrogen, ammonia** and **nitrates**. Nitrogen is *fixed* from the **atmosphere** by:

(a) The **Haber process**.
(b) Electrical discharges (lightning).
(c) Action by soil bacteria.

The nitrates in the soil are taken up by plants which are then eaten by animals. Both animal and plant **protein** can rot and revert to ammonia. Animals also excrete urine which decomposes to ammonia.

Nitrogen is returned to the atmosphere by the action of other bacteria.

Despite human extraction of **nitrogen** from the air in the Haber process the amount of nitrogen in the atmosphere remains approximately constant.

Nitrogen oxides

Nitrogen monoxide NO	This is a colourless gas which can be produced by the action of *moderately* concentrated **nitric acid** on **copper**. The gas reacts *immediately* with oxygen to form nitrogen dioxide: $$2NO(g) + O_2(g) \rightarrow 2NO_2(g)$$
Nitrogen dioxide NO_2	This is a brown gas. It has a choking smell and irritates the lungs and windpipe. Breathing it can lead to death from pneumonia. It can be produced in the laboratory by the

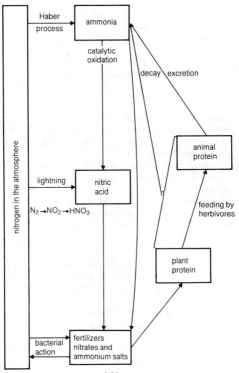

161

| | action of concentrated **nitric acid** on **copper**. |
| Nitrous oxide N_2O | This is a colourless gas with a sweetish smell. It is an anaesthetic and is used in dentistry. It is sometimes called laughing gas. |

Noble gases (or **Inert gases**)   These elements are in **group** o of the **periodic table**. They are all very unreactive **monatomic** gases which occur in the **atmosphere** to small extents:

Helium, Neon, Argon, Krypton and Xenon.

Their **electronic configurations** are very stable and the elements show no tendency to lose or gain electrons. Because of this, they find it hard to form compounds. Compounds do exist of the more massive elements, e.g. XeF_4, but these are rare.

Noble gas structure (or **Inert gas structure**)   Noble gases are stable, unreactive elements because of their **electronic configurations**. They have *eight* electrons in their *outer* shell. This is a stable configuration because the eight electrons completely fill a part of the shell making it difficult to add another or take one away.

When elements form ions, the ions which are formed have the inert gas configuration:

Element	Ion	Electronic configuration		Noble gas
		atom	ion	
Sodium	Na^+	2·8·1	2·8	Neon
Fluorine	F^-	2·7	2·8	Neon
Calcium	Ca^{2+}	2·8·8·2	2·8·8	Argon
Sulphur	S^{2-}	2·8·6	2·8·8	Argon
Aluminium	Al^{3+}	2·8·3	2·8	Neon
Bromine	Br^-	2·8·18·7	2·8·18·8	Krypton

Nonmetal Elements which have (a) either molecular structures, and thus are **gases at room temperature**, or are **solids** or **liquids** with low **melting** and **boiling points**; or (b) **giant structures** with **covalent bonding**. Typical properties of nonmetals are:

> Poor **conductivity**
> Do not react with **acids**
> Produce **acidic oxides**
> Form **covalent** compounds
> Give rise to **anions**

Examples:
The halogens, the noble gases, oxygen, sulphur, carbon, nitrogen.

NTP See **STP**.

Nuclear reactions Nuclear reactions are very different from normal chemical reactions. New **compounds** are not formed but changes occur in the nuclei and new *elements* can be formed. Most elements have stable and unstable isotopes, e.g.

carbon-12 carbon-13   carbon-14

     stable   unstable

But some have no stable isotopes, e.g. **uranium, plutonium**. Whether an isotope is unstable depends on the numbers of **neutrons** and **protons** in the **nucleus**. When an isotope is unstable, several reactions can occur:

(a) The isotope can *split* — fission can occur. This creates two stable isotopes. Neutrons are also released and lots of energy is released.

(b) A neutron can decay into a proton and an **electron**:

$$_0^1n \rightarrow {}_1^1p + {}_{-1}^0e$$

The electron is expelled from the nucleus (**beta particle**) and the **atomic number** of the atom *increases* by 1, e.g.:

$$_{82}^{209}Pb \rightarrow {}_{83}^{209}Bi + {}_{-1}^0e$$

This creates a new element. There is no change in mass.

(c) An **alpha particle** is expelled from the atom, e.g.:

$$_{92}^{238}U \rightarrow {}_{90}^{234}Th + {}_2^4He$$

(d) An unstable atom can emit **gamma rays**. This often occurs after the emission of a beta particle. See **Half-life**.

These nuclear reactions have been used by us in many ways. Nuclear power stations use the heat produced in the fission of uranium, plutonium or thorium isotopes. The heat raises steam which drives a turbine producing electricity. This same source of energy is also used by some countries in the form of bombs. These have terrible destructive powers.

Isotopes producing gamma rays are used to destroy bacteria in food processing and *cancerous* cells in the body.

Isotopes are also used in industry for a variety of analytical purposes, e.g. detecting cracks in pipelines.

Scientists are striving to produce useful energy from **fusion** reactions and if this is possible it will provide a source of **electricity** which is not dependent upon **coal** or **petroleum**.

The products of radioactive decay (**radioactivity**) are dangerous as they can produce harmful effects on the body including *leukaemia* and *cancers*. Great care has to be taken when using radioactive materials.

Nucleus   This is the part of an **atom** where the **mass** is concentrated. It contains **protons** and **neutrons** and is usually pictured as being the compact centre of a spherical atom.

Electrons move *around* the nucleus. The nucleus has a positive charge and in the neutral atom this is balanced by the charges on the electrons.

Hydrogen-1 **isotopes** are the only atoms whose

nuclei contain *no* neutrons. See diagram below. See also **Nuclear reactions**.

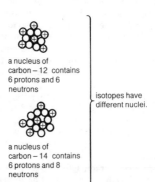

a nucleus of carbon – 12 contains 6 protons and 6 neutrons

a nucleus of carbon – 14 contains 6 protons and 8 neutrons

isotopes have different nuclei.

Nylon The nylons are a family of **polyamide polymers**. The most common of them is *Nylon 6.6*. This *man-made fibre* has great strength. Its uses include fabrics (shirts, cloth), yarns (stockings, knitwear), carpets, ropes and nets.

Nylon is useful because it will not rot, it does not absorb water but it does stretch. This is useful in ropes and stockings. It is often mixed with other fibres, e.g. wool, to get the correct balance of properties.

6 carbon atoms   6 carbon atoms

Nylon 6.6

Octane rating **Petrol** and **air** mixtures have to explode at exactly the correct moment in an internal combustion engine. If the wrong kind of fuel is used ignition of the mixture can occur *before* it should. This leads to the characteristic sound of *pinking* (or *knocking*).

The octane rating (or octane *number*) of a petrol is a measure of the *anti-knock* qualities of the fuel. The higher the rating, the better it is. Four star petrol has an octane rating of 98. The greater the proportion of *branched* **hydrocarbon** molecules in the fuel, the greater the rating will be.

Octane ratings can be raised by adding substances such as tetraethyl lead. This unfortunately leads to lead pollution in the atmosphere.

Oil A liquid fat, e.g. melted butter, olive oil, sunflower oil, etc. The word is sometimes used in place of **petroleum**.

Oleum A solution of **sulphur(VI)** oxide in *concentrated* **sulphuric acid**. It is a very corrosive substance and is a vigorous **oxidizing agent**. See **Contact process**.

Ore A naturally occurring substance from which an **element** can be extracted, e.g.:
galena — lead, haematite — iron, calomine — zinc.

Organic chemistry The study of **compounds** of **carbon**. It does not include carbonates, carbon dioxide, etc. See **Inorganic chemistry**.

Oxidation A substance undergoes oxidation if it:

gains oxygen: $2Mg(s) + O_2(g) \rightarrow 2MgO(s)$
loses hydrogen: $CH_4(g) + Cl_2(g) \rightarrow$
$CH_3Cl(l) + HCl(g)$
loses electrons: $Cu(s) \rightarrow Cu^{2+}(aq) + 2e$

See **Oxidizing agent, Reduction** and **Redox**.

Oxide A compound formed between an **element** and **oxygen** only. Oxides can be formed by:
  (a) direct combustion of the elements
  (b) oxidizing compounds
  (c) the action of heat on a **carbonate**, an **hydroxide** or some **nitrates**.

Examples:
$2Mg(s) + O_2(g) \rightarrow 2MgO(s)$
$CH_4(g) + 2O_2(g) \rightarrow CO_2(g) + 2H_2O(g)$
$CuCO_3(s) \rightarrow CuO(s) + CO_2(g)$
$Pb(OH)_2(s) \rightarrow PbO(s) + H_2O(g)$
$2Zn(NO_3)_2(s) \rightarrow 2ZnO(s) + 4NO_2(g) + O_2(g)$

Oxides can be reduced to the element with a

suitable **reducing agent**, e.g. hydrogen, carbon or carbon monoxide. See **Basic oxide**, **Acidic oxide**, **Amphoteric oxide**.

Oxidizing agent Something which causes the **oxidation** of another substance. When this happens the oxidizing agent is reduced.

Common oxidizing agents:

Oxygen	O_2
Chlorine	Cl_2
Ozone	O_3
Hydrogen peroxide	H_2O_2
Potassium manganate(VII)	$KMnO_4$

Oxidizing agents at work:
$$Mg(s) + O_2(g) \rightarrow 2MgO(s)$$
$$Cl_2(g) + 2Na(s) \rightarrow 2NaCl(s)$$
$$H_2O_2(aq) + H_2S(g) \rightarrow 2H_2O(l) + S(s)$$

Oxonium ion (H_3O^+) The loss of an **electron** from a **hydrogen atom** leads to the formation of a hydrogen ion. This is a proton.

The proton is such a small, reactive particle that in **aqueous** solution it forms bonds to water molecules. More than one water molecule is probably involved but, for simplicity only one is usually shown.

$$H \rightarrow H^+ + e^-$$

$$H^+ \cdots\cdots O \begin{matrix} H \\ \\ H \end{matrix}$$

$$H_3O^+(aq)$$

169

It can be used in ionic equations in place of $H^+(aq)$:

$$H_3O^+(a) + OH^-(aq) \rightarrow 2H_2O(l)$$
$$\text{acid} + \text{alkali}$$

Oxygen (O_2) A gaseous, nonmetallic **element** in **group** VI of the **periodic table**. It makes up 21% of the **atmosphere**. It is a vigorous **oxidizing agent** and is vital for the **respiration** of plants and animals.

It is a reactive gas, readily forming **oxides** with most **elements**. It is obtained industrially by the **fractional distillation** of liquefied air and in the laboratory by the decomposition of **hydrogen peroxide**:

$$2H_2O_2(aq) \rightarrow 2H_2O(l) + O_2(g)$$

A **catalyst** such as manganese(IV) oxide is usually used.

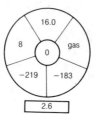

Oxygen is a colourless, odourless **diatomic** gas. It is neutral and is only slightly soluble in water ($40\ cm^3$ per cubic decimetre). It is sufficiently soluble to allow fish and other aquatic life to live in water. Problems arise if the oxygen in water drops to too low a level. Then, fish suffocate and die. Such a drop in the **concentration** of oxygen is usually caused by pollution.

The chemical test for oxygen is to plunge a glowing splint into the gas. If the gas is oxygen, the splint will relight. The gas will also make anything which is already burning in air, burn much more fiercely.

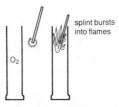

splint bursts into flames

O_2

Oxygen is used extensively in steel-making and welding, as a rocket propellant together with kerosene or hydrogen, and for life-support systems in medicine or breathing apparatus.

Ozone (O_3) This is an **allotrope** of **oxygen** where the molecule is *tri*atomic $O_3(g)$. It is a blue gas and is a very powerful **oxidizing agent** and extremely poisonous. It occupies a very small part of the **atmosphere** and is mainly located about 30 km above the surface — the *ozone layer*. Here it prevents most of the **ultraviolet radiation** from reaching the earth where it could do great damage. It can be formed from oxygen by passing a spark through the gas:

$$3O_2(g) \rightarrow 2O_3(g)$$

Ozone is a powerful germicide, and its great oxidizing power makes it useful, in very high dilution, for ventilating spaces where fresh air has limited access, e.g. underground railways.

Paraffin This is a fuel which is obtained from **petroleum**. It has a boiling range of 160−250°C. See **Kerosine**.

The term was once used instead of **alkane**.

Paraffin wax consists of a mixture of solid **hydrocarbons**. It is used to make candles.

Pascal This is the **SI unit** of **pressure**. One **atmosphere** pressure is about one hundred kilopascals, (100 kPa). One pascal is equivalent to a force of one Newton on one square metre (1 Nm^{-2}).

Passive Describes a metal which posseses a surface layer of **oxide** which makes it unreactive − passive. **Aluminium** has such a layer and **iron** can be given one by dipping the metal into concentrated **nitric acid**.

Peptide When two or more **amino acids** react together, peptides are formed. If three or more amino acids are involved the term **polypeptide** is usually used. **Proteins** consist of long polypeptide chains which are often linked together in a variety of ways.

Peptides contain the *peptide link:*

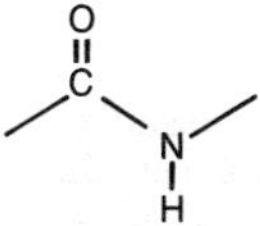

Percentage composition In analysis it is often useful to know what proportion of a compound is made up of any given element. This is usually expressed in percentage terms:

Examples	% by *mass* of	
Methane	carbon	hydrogen
CH_4	75%	25%
Ethyne C_2H_2	92%	8%

Knowing the percentage composition of a compound it is possible to work out its **empirical formula**.

Example:
A compound is known to have a percentage composition of: Ca 40% C 12% O 48%.

To find the *molar* ratio of the elements in the compound, divide each figure by the respective A_r value (Ca) $^{40}/_{40} = 1$ (C) $^{12}/_{12} = 1$ (O) $^{48}/_{16} = 3$.

The molar ratio is: 1 : 1 : 3.

Therefore the ratio of the atoms is: 1 : 1 : 3.

The empirical formula is: $CaCO_3$.

Period In the **periodic table** the horizontal rows of **elements** are known as periods. There are seven in all. See **Group**.

Periodic table The periodic table which is shown on page 176 is a way of presenting all the **elements** so as to show their similarities and differences.

The elements are arranged in increasing order of **atomic number (Z)** as you go from left to right across the table.

Period 3							
Na	Mg	Al	Si	P	S	Cl	Ar
11	12	13	14	15	16	17	18

The horizontal rows are called **periods** and the vertical rows, **groups**. A **noble gas** is found at the right hand side of each period.

Period	1	He
	2	Ne
	3	Ar
	4	Kr
	5	Xe
	6	Rn

There is a progression from metals to nonmetals across each period.

Similar elements are found in a group, e.g. these

elements have a similar **electronic configuration** and
are found in the same group.

group I	alkali metals
	Li Na K
group VII	halogens
	F Cl Br I

The number of **electrons** in the outer shell is the same
as the number of the group, e.g. group I.

lithium	2·1
sodium	2·8·1
potassium	2·8·8·1

The block of elements between groups II and III are
called the **transition metals**. These are similar in many
ways e.g. they produce coloured compounds, have
variable valency and are often used as catalysts.
Elements 58 to 71 are known as *Lanthanide* or *Rare
Earth* elements. These elements are found on earth in
only very small amounts.

Elements 90 to 103 are known as the *Actinide*
elements. They include most of the well known
elements which are found in **nuclear reactions**. The
elements with larger atomic numbers than 92 do not
occur naturally. They have all been produced
artificially by bombarding other elements with
particles. **Plutonium** is formed in nuclear reactors.

Elements with atomic numbers of 104 and above
are not being named after famous scientists or places.

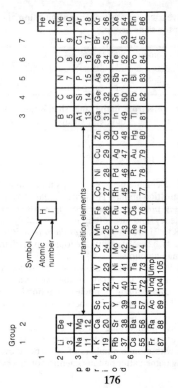

They are being named in a systematic way according
to their atomic number:

element 104 is *Unnilquadium*	Un = 1	Symbol
	nil = 0	*Unq*
	quad = 4	
element 118 would be *Ununoctium*	un = 1	
	un = 1	*Uuo*
	oct = 8	

Peroxide A peroxide contains the O_2^{2-} **ion** or $-O_2$
group. The best example is **hydrogen peroxide**
(H_2O_2). The peroxides of alkali metals are also
known. Hydrogen peroxide can be made by the action
of acid on these:

$$Na_2O_2(s) + 2HCl(aq) \rightarrow 2NaCl(aq) + H_2O_2(aq)$$
$$\text{or } O_2^{2-}(s) + 2H^+(aq) \rightarrow H_2O_2(aq)$$

Hydrogen peroxide can be used in dilute solution as a
disinfectant and a bleach. See **Hydrogen peroxide**.
Peroxides are vigorous **oxidizing agents.**

Petrochemical This is a chemical which has been
made from petroleum. Examples would include
ethene and propene. These chemicals are used in the
manufacture of other chemicals, e.g. **poly(ethene)**
and **poly(propene)**. Petrochemicals are *intermediates*
in the production of finished products.

Petrol This is a mixture of **hydrocarbons** which is
used as a fuel in internal combustion engines. It is

produced from **petroleum** in a refinery. It is principally a mixture of $C_5 - C_{10}$ alkanes obtained by both straight fractional distillation and **cracking** and **reforming**. See **Octane rating**. Petrol is also known as gasoline or motor spirit.

Petroleum   This is the term describing the mixture of **hydrocarbons** which are found in the earth's crust, e.g. **natural gas** and *crude oil*. It has been produced over millions of years from the remains of *marine* animals and plant organisms. It is the raw material for the petrochemical industry and is the source of our **petrol, diesel** fuel, heating oil, fuel oil and gas supplies. Vast reserves of petroleum are found in the Middle East, the United States, the Soviet Union, Central America and the North Sea.

Petroleum has replaced **coal** as the chief source of raw materials for the chemical industry. But the supply of petroleum will not last forever and the search is now on for the substance which will replace it in our lives.

How petroleum is converted into useful products is described in **refining**. See also **Fractional distillation**.

Pewter   An **alloy** which was made of **lead** and **tin** (80–90%) with small amounts of antimony. It was widely used for plates and mugs and is now used as jewellery. The pewter made now contains no lead because of the danger of the lead getting into solution or contaminating food. It has been replaced by **copper** and antimony.

pH

acidity increasing		neutral	alkalinity increasing	
0	3	7	10	14

The pH scale is a measure of the acidity or *alkalinity* of a solution. The lower the value, the more acidic is the solution, i.e. the larger the **concentration** of **oxonium ions** there are within it. A **neutral** solution, where the concentrations of oxonium and hydroxide ions are equal, has a pH of 7 at 25°C.

$pH = -\log_{10}[H_3O^+]$
The pH of a solution is the negative logarithm (base 10) of the concentration of oxonium ions (mol/dm^3)

Example:
The pH of a solution whose concentration is 0.1 mol/dm^3 is $-\log(0.1) = 1$

pHs of common substances	
substance at a concentration of 1 mol/dm^3	pH
strong acid (HCl)	0
weak acid (CH$_3$COOH)	4
water	7
ammonia solution	10
strong alkali (NaOH)	14

Phase change (or **change of state**)   A phase change occurs when a substance goes from one physical state to another. The possible phases in a chemical system are shown below:

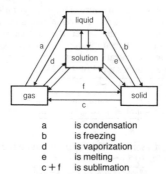

a	is condensation
b	is freezing
d	is vaporization
e	is melting
c + f	is sublimation

Phenol   A colourless crystalline **aromatic** solid, which turns pink on exposure to air and light.

C_6H_5OH

It is a very corrosive chemical and is poisonous. It is useful as an antiseptic, and **disinfectant** and many household products are based on similar molecules to phenol, i.e. compounds with similar (or better) disinfectant properties but which are less corrosive, e.g. T.C.P. and Dettol.

The O — H group in phenol has an acidic hydrogen and salts can be formed:

$$C_6H_5O^-Na^+(s)$$

Phenolphthalein This is an **indicator** which is used to follow **acid-base** reactions. The molecular structure of the compound is based on the **phenol** molecule.

acidic solution	alkaline solution
← colourless	red →

Phosphates These are **salts** of **phosphoric(v) acid**. Phosphoric acid is a *tri*basic acid and, therefore, gives rise to three kinds of salts. These are shown here:

sodium phosphate	Na_3PO_4
sodium hydrogenphosphate	Na_2HPO_4
sodium dihydrogenphosphate	NaH_2PO_4

Phosphates are used in fertilizers to replace the **phosphorus** — containing compounds in the soil. It comes in several forms:

Superphosphate — a mixture of calcium sulphate and calcium dihydrogenphosphate $(Ca(H_2PO_4)_2)$.

A mixture of ammonium nitrate and ammonium hydrogenphosphate $(NH_4)_2HPO_4$ is made by reacting **ammonia** with a mixture of **phosphoric(v)** and **nitric acids**.

181

Calcium phosphate is the chief constituent of animal bones and phosphates are used extensively in washing powders and detergents.

Slag from **blast furnaces** is put directly onto the soil since it contains calcium phosphate.

Phosphoric(v) acid (H_3PO_4) A tribasic acid which gives rise to three kinds of **phosphate**. It is a **solid** at **room temperature** but is usually sold as a viscous **solution** in water. It is used for rust-proofing steel by forming a protective layer of iron phosphate. It is also used in the food and drug industries.

Phosphorus (P_4) Phosphorus is a solid nonmetallic **element** in **group** v of the **periodic table**. Three **allotropes** exist: white, red and black phosphorus. The information in the chart refers to *white* phosphorus.

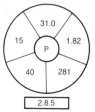

The allotropes have very different physical properties, e.g. **crystal** structure, **density, melting** and **boiling points**.

White phosphorus is very poisonous and has to be kept under water because it bursts into flame when it comes into contact with air, forming the oxide:

$$P_4(s) + 5O_2(g) \rightarrow P_4O_{10}(s)$$

Rocks containing phosphorus compounds are quite common in the earth's crust and phosphorus is an important element for the maintenance of life. Plants need it and so it is a necessary component of fertilizers e.g. superphosphate. See **Phosphates**. Some enzymes used in **respiration** also contain the element. Animal bones contain **phosphates**. In earlier days phosphorus was used extensively in making the heads of matches. It has since been banned. Matches nowadays contain phosphorus sulphide, and indeed in 'safety matches' the phosphorus is located on the side of the box, hence the match will only strike on the box.

Photochemical reactions Chemical reactions which are brought about by light. The process of photography depends on light. When silver bromide is exposed to light it decomposes:

$$2AgBr(s) \rightarrow 2Ag(s) + Br_2(g)$$

The black part of the negative of a film is a thin layer of **silver**. This is then processed to produce the final pictures.

Photosynthesis is the most important photochemical reaction. Our existence on this planet depends on it.

Other examples are the reaction between **hydrogen** and **chlorine** which can be started off by a bright light e.g. burning magnesium.

$$H_2(g) + Cl_2(g) \rightarrow 2HCl(g)$$

In hot areas of the world, e.g. California, where there are a lot of vehicle exhaust fumes, photochemical *smog* can be produced by the reaction of the exhaust fumes. This is an acrid haze which is very unpleasant.

The **bleaching** action of the sun is also a photochemical reaction as is the conversion of cholesterol in the skin into **vitamin** D.

In all these reactions light provides the **energy** that makes the reactions work.

Photosynthesis A **photochemical** process in which **carbon dioxide** and **water** are converted into **carbohydrate** and **oxygen**:

$$6CO_2(g) + 6H_2O(l) \rightarrow C_6H_{12}O_6(s) + 6O_2(g)$$

This occurs in the leaves of plants and is the means whereby plants obtain their *food*. It also puts **oxygen** back into the **atmosphere**.

The reaction is catalysed by **chlorophyll** and sunlight provides the **energy** for the reaction.

Physical change A physical change to a substance involves changes in its physical **properties** with no alterations to chemical properties.

Physical chemistry   This is the study of the physical **properties** of elements and compounds and the relationship between their chemical properties and physical properties.

Pipette   A piece of glassware used for measuring and transferring a fixed **volume** of liquid, e.g. 25 cm^3.

s

graduation
mark on
the glass

jet

The liquid is drawn up into the pipette through the jet by suction applied at s. Liquid is taken up until the bottom of the meniscus is on the graduation.

To remove the liquid from the pipette, the suction is released and the jet is touched against the side of the vessel into which the liquid flows. In these circumstances, the pipette will release exactly the correct volume of liquid.

A safety pipette filler should always be used to fill a pipette. See **Burette**.

Planar   A planar molecule is one which has all its atoms in the same plane, i.e. it is flat. All **linear** and *tri*atomic molecules must be planar. Boron trifluoride is planar.

Other planar molecules include:

H_2O	C_2H_2	CO	HCl
CO_2	SiO_2	NO_2	N_2

Most carbon compounds are *nonplanar* because of the **tetrahedral** arrangement of atoms around the carbon atom. Most alkenes are nonplanar but **ethene** is an exception.

Propene is *nonplanar* because the three hydrogen atoms attached to the carbon atom are above or below the other atoms.

ethene is planar   but propene is not

Plastic Capable of being shaped or moulded by **heat** and **pressure**. The word is usually used to describe man-made **polymers**, e.g. **poly(ethene), nylon**, etc. Over thirty different types of polymers exist and by combining them it is now possible to produce a man-made substance with specific properties. See **Thermosetting, Thermoplastic**.

Platinum Platinum is a **transition metal** which is used as a **catalyst** in the chemical industry. It is also used as a metal for jewellery and as inert electrodes in electrolysis. It is expensive because it is useful and rare. Platinum occurs in nature as the element.

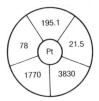

Plutonium A man-made element. It is made in nuclear reactors from the 238 **isotope** of **uranium** by the absorption of a **neutron**:

$$^{238}_{92}U + ^{1}_{0}n \rightarrow ^{239}_{92}U \text{ this is unstable}$$

$$^{239}_{92}U \rightarrow ^{239}_{93}Np + ^{0}_{-1}e \text{ (beta particle)}$$

Neptunium−239 is also unstable:

$$^{239}_{93}Np \rightarrow ^{239}_{94}Pu + ^{0}_{-1}e \text{ (beta particle)}$$

Uranium−239 and neptunium−239 have **half-lives** of twenty minutes and two days respectively, and so soon decay. The half-life of plutonium−239 is over 24 000 years and so is fairly stable.

Plutonium is used in bombs and as a fuel for nuclear reactors, e.g. *fast breeder reactors*. The cores of these

187

reactors are surrounded by uranium—238 which is connected to plutonium—239 (in the way shown above) while the reactor is working. In this way it is possible to produce more fuel than is used in the reactor.

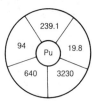

Plutonium is a toxic and dangerous chemical and must be handled with great care. Great controversies surround its use.

Poison A poison is a substance which is harmful if taken into the body and which will lead to death or injury. Poisons can be breathed in, e.g. hydrogen cyanide (HCN), **carbon monoxide** (CO) or lead fumes in the air; they may be eaten or drunk, e.g. arsenic(III) oxide, strychnine, or paraquat or they may pass into the body through the skin, e.g. **benzene** and nerve gases.

Radiation from nuclear **isotopes** can also be viewed as a poison.

Polyamides Man-made **polymers** which contain the

amide group of atoms. Some compounds are usually known by the name **nylon**. Polyamides are **condensation** polymers.

amide group

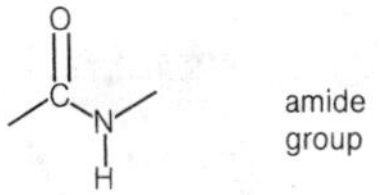

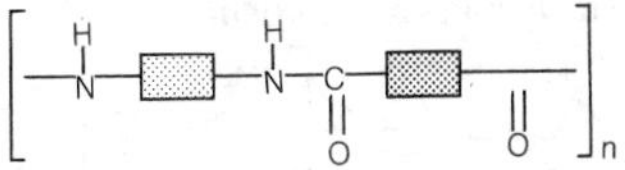

Poly(chloroethene) (PVC) This **polymer** is made from chloroethene ($CH_2 = CHCl$). PVC has a wide range of uses: vinyl flooring, food containers, records, fibres, guttering.

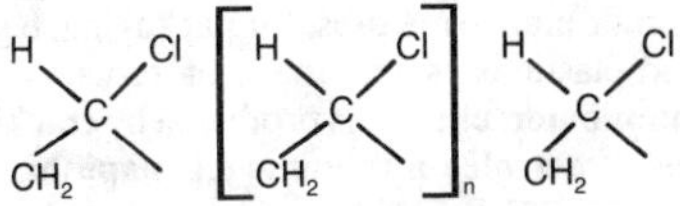

Polyesters Man-made **polymers** which contain an **ester** group of atoms.

A well-known polyester is **Terylene** which is widely used in everyday life. They are *condensation* polymers whose **monomers** are an **acid** and an **alcohol**.

189

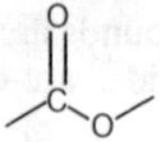

ester
group

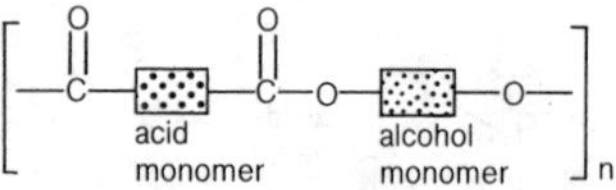

Poly(ethene) This **polymer** is made from **ethene** and is usually known as *polythene*.

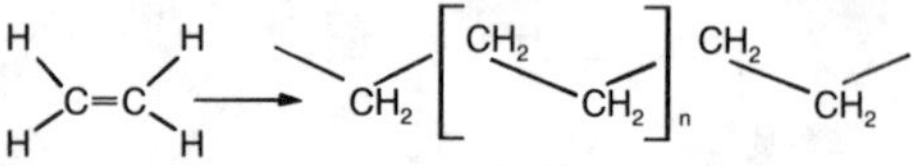

$(C_2H_4)n$

Poly(ethene) is an **addition polymer**. Here, n is a very large number (about 50 000). Poly(ethene) is widely used today. The polymer is a **saturated alkane** and so is very unreactive. It is used for containers, e.g. ice cream or margarine tubs, for packaging, e.g. cling films and plastic bags, and many other uses.

The **monomer**, ethene, is produced by **cracking** the products of **petroleum** refining, e.g. **naphtha**.

Polymers Polymers are large molecules in which a group of atoms are repeated, e.g.

$$x - x - x - x - x - x - x - x$$

or

$$x - y - x - y - x - y - x - y$$

190

Polymers are either naturally occurring or man-made. Starch and cellulose are natural ones whilst **nylon, Terylene, poly(chloroethene)** and **poly(propene)** are man-made.

Polymers are made by reacting **monomer** molecules together usually with a **catalyst**.

Addition polymerization involves one kind of unsaturated molecule. In the **polymerization** a long chain of atoms (linked by **single bonds**) is formed. See **poly(ethene)**. Because the polymer is a **saturated** molecule it is usually fairly *un*reactive and this gives the polymer useful properties. See the entries for individual polymers.

Example:

monomer (acrylonitrile)

polymer (used to make fabrics such as *Acrilan* and *Courtelle*).

Condensation polymerization involves two molecules (the monomers) which condense together into long chains. A small molecule is eliminated during the reaction. The monomers are difunctional molecules, i.e. each molecule possesses two functional groups. **Nylon** and **Terylene** are good examples of condensation polymers. Nylon is made by reacting an acid and an amine:

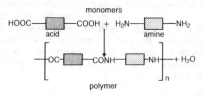

The source of most of the raw materials used in making man-made polymers is **petroleum**. There has been great growth in the industry in recent years and will no doubt continue as long as petroleum or a substitute is available.

Polymerization This is the reaction of **monomers** to form a **polymer**.

Polypeptide A polypeptide is a molecule which is made up of up to 50 **amino acids** linked together with peptide links. If there are more than 50 amino acids the molecule is termed a **protein**. Polypeptides are important molecules in the body, e.g. acting as **enzymes**.

Poly(phenylethene) (or polystyrene) This is a **poly-**

mer based on the **alkene** phenylethene. It is an addition polymer and is used widely in the form of sheets mainly, in an *expanded* form. Here **air** is blown into the polymer and the result is a white granular solid which has excellent insulation properties. It is used for cups, ceiling tiles and other items which need to be poor **conductors.**

Phenylethene C_8H_8

Poly(propene) This **polymer** is similar to **poly(ethene)** in that it is an addition polymer produced from an **alkene**. The alkene is propene (C_3H_6) which is obtained from petroleum. The polymer is versatile and useful. It is strong and hard wearing and finds use as carpet fibres and material for laboratory flasks and beakers. It is very unreactive. It is also known as polypropylene.

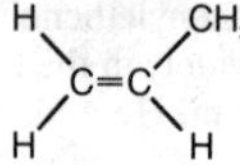

Propene C_3H_6

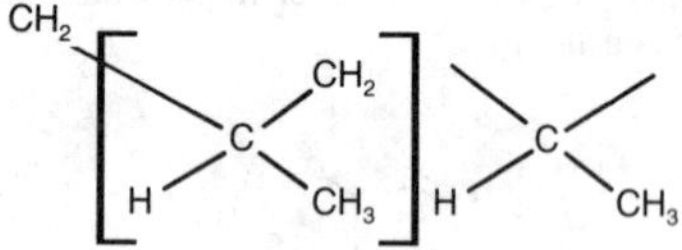

poly(propene)

Polysaccharides These are **polymers** made up of **sugar** molecules such as **glucose**. There are three common naturally occurring examples:

$\left.\begin{array}{l}\textbf{starch} \\ \textbf{cellulose} \\ \text{glycogen}\end{array}\right\}$ These three are all made up of glucose molecules in chains.

Polysaccharides can be broken down by **hydrolysis** into simple sugars such as glucose or into **disaccharides** such as **maltose**. Hydrolysis is brought about by **enzymes** in plants and animals but can also be done by the use of acids.

Polysaccharides are valuable foods, e.g. starch in the form of potatoes or rice; cellulose is found in all plants and glycogen is the form in which excess **carbohydrate** is stored in animals.

Poly(tetrafluoroethene) (PTFE) *Poly*tetra*fluoro-*

ethene. This **polymer** is *very* inert. It is the non-stick surface for saucepans, e.g. *Teflon, Fluon.* It is also used for bearings because it has a low coefficient of friction.

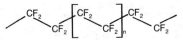

Polythene See **Poly(ethene)**. The older term 'polythene' is still used in industry and shops.

Potassium Potassium is a **group** I metal. It is a grey, very soft element which is very reactive. It is easily cut with a knife revealing a silvery surface which tarnishes immediately. It is a powerful **reducing agent**, giving rise to the potassium ion (K^+):

$$2K(s) + 2H_2O(l) \rightarrow 2KOH(aq) + H_2(g)$$

It is stored under oil because of its reactivity towards air and water. The metal ion gives a lilac flame test.

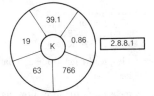

Potassium is obtained by the **electrolysis** of molten potassium chloride (KCl).

Potassium compounds These tend to be colourless, crystalline compounds which are very soluble in water. Potassium **salts** play an important role in the body:

Potassium iodide KI	The solution is a useful solvent for **iodine** forming KI_3(aq). This is used to test for **starch**.
Potassium nitrate KNO_3	*Saltpetre*. This is used in food preservation and in explosives.
Potassium manganate(VII) $KMnO_4$	This is a vivid purple crystalline compound which is a good **oxidizing agent**.

ppt This is an abbreviation for **precipitate**.

Precipitate (ppt) The insoluble substance formed on mixing the two solutions in a **double decomposition** reaction is called a 'precipitate'.

E.g.

$$Pb(NO_3)_2(aq) + 2NaCl(aq) \rightarrow$$
$$PbCl_2(s) + NaNO_3(aq)$$
$$BaCl_2(aq) + Na_2SO_4(aq) \rightarrow$$
$$BaSO_4(s) + 2NaCl(aq)$$

In these examples, lead(II) chloride and barium sulphate are the precipitates.

Pressure A measure of the force pressing onto an object's surface. The **SI unit** of pressure in the **pascal**. Other units used are **atmospheres** and **mmHg**.

The pressure that a gas exerts upon its container is caused by molecules striking the container's walls. The pressure of a gas depends upon its volume and its temperature. See **Boyle's law**, **Charles' law** and **Gas laws**.

Proof A measure of how much **ethanol** there is in a solution of ethanol in water. Spirits such as whisky are normal sold as *70° proof*. This is a 40% solution of ethanol by volume.

Propane An **Alkane** which is obtained from **petroleum**. It is chiefly used as a fuel:

$$C_3H_8(g) + 5O_2(g) \rightarrow 3CO_2(g) + 4H_2O(g)$$

Although a gas at **room temperature**, it is easily liquified and it is sold as bottled gas, e.g. *Calor Gas* or *Camping Gaz* where it is mixed with **butane**.

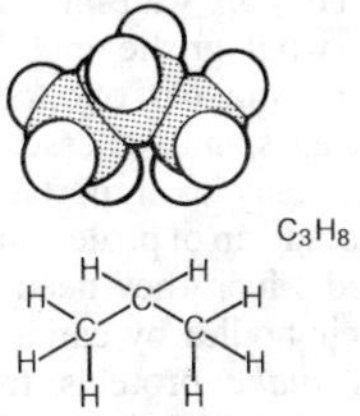

C_3H_8

Properties The properties of a substance are those characteristic ways in which it behaves (reacts) that

197

make it what it is, and make it different from other substances.

It is usual to classify properties as *physical* or *chemical.* Chemical properties are concerned with the substances reactions.

Physical properties include:	Chemical properties include: whether the substance
colour	is a metal or nonmetal;
density	gives acidic or basic oxides;
physical state	has more than one valency;
boiling point	reacts with acids;
melting point	is an oxidizing or reducing agent
crystal form	
solubility	
hardness	

Proteins Large, **organic** compounds made up of chains of **amino acids**. The amino acids are joined by **peptide** links. They are widespread in animal bodies: skin, hair, nail, wool, gristle, muscle.

Proteins form a vital part of our diet. Good sources are: meat, fish, eggs, milk, cheese, beans, bread. The **enzymes** which play so important a role in living organisms are made up of protein molecules. Animals obtain the food which they need in order to make proteins in their bodies by eating other animals or plants. Plants *make* proteins from the nitrogen containing compounds in the soil. See **Fertilizer**.

Proton A positively charged subatomic particle found in the **nucleus** of the **atom**. The number of

protons in an atom is the same as the **atomic number.**
Isotopes of any element *always* contain the same
number of protons.

PTFE   See **Poly(tetrafluoroethene).**

Pure   A substance is pure if it contains only one
element or **compound**, e.g. sodium chloride crystals
can be obtained in a pure state, but table salt impure
because it is mixed with other substances.

PVC (polyvinylchloride)   See **Poly(chloroethene).**

Pyrolysis   The decomposition of a substance by the
action of heat:

$$CaCO_3(s) \rightarrow CaO(s) + CO_2(g)$$

Qualitative   Qualitative statements or analysis are
concerned with composition, *not* with amounts: e.g.
Water is a compound of hydrogen and oxygen.

Quantitative   Quantitative statements or analyses
are concerned with amounts: e.g.
Water consists of two atoms of hydrogen and one of
oxygen.

Radical   Refers to inorganic chemistry. It is the atom
or group of atoms present in a compound which is
responsible for the characteristic properties of that
compound. E.g.

$-SO_4^{2-}$	sulphate
$-CO_3^{2-}$	carbonate
$-OH^-$	hydroxide

Radioactivity This is the spontaneous disintegration of the nucleus of an atom accompanied by the emission of electromagnetic radiation (γ rays) or particles (α or ß particles).

Not all elements have radioactive isotopes, though many can now be created artificially. Some radioactivity occurs naturally, e.g. isotopes found within rocks. Granite usually has quite a large concentration. Living things contain radioactive carbon and this can be used for finding the age of their remains (see **Radiocarbon dating**). Types of emissions which accompany radioactive decay include:

Electrons (beta particles) can be emitted.

Alpha particles can be emitted.

Gamma rays can be emitted.

See **Nuclear reactions** and **Half-life**.

Radiocarbon dating Carbon-14 (^{14}C) is a naturally occurring radioactive isotope. All living things contain ^{14}C, kept at a constant level by continuous exchange through feeding and respiration. This exchange ceases on death and the ^{14}C level falls at a constant rate due to radioactive decay. Scientists measure the amount of radioactivity left in remains, and from this can calculate the age of the remains.

Rate of reaction The rate (or speed) at which a reaction occurs depends on several factors:

Temperature: The higher the temperature, the

200

faster the reaction is because the particles are moving with greater energies.

Particle size: The smaller the particles involved, the greater the surface area where the reaction can take place and the faster is the reaction.

Concentration: The more concentrated a solution (or the higher the **pressure** of a gas) the faster a reaction will occur. This is because there are more particles in a given volume to react.

Catalysts: These change the rate by providing an alternative reaction pathway along which the reaction can occur.

Rayon This was one of the first man-made fibres. It is made from **cellulose** (from wood). Wood contains cellulose fibres but they are too short to spin into yarn. Cellulose is dissolved in sodium hydroxide/carbon disulphide solution and squirted into dilute **sulphuric acid**. Here the cellulose is reformed as long threads which are ready for weaving. Rayon is used to make clothes and curtaining.

Reaction When a chemical reaction occurs a new substance is formed. The new substance will have different properties from the reactants.

Reactions can be accompanied by: heat, light, sound, colour changes.

Reactivity series This is a series of elements arranged in order of their chemical reactivity.

Such a series is shown here:

↑ Potassium
Sodium
Calcium
Magnesium
Aluminium
Zinc
Iron
Lead
Copper
Silver

The series shown here is a short one containing only a few elements but the same general conclusions apply: **Oxides** will be *reduced* by **elements** above them but not the other way round:

$$Mg(s) + CuO(s) \rightarrow MgO(s) + Cu(s)$$
$$Cu(s) + MgO(s) - \text{no reaction}$$
$$Cu(s) + ZnO(s) - \text{no reaction}$$

Metal ions will be *reduced* by elements above them but not the other way round:

$$Zn(s) + 2Ag^+(aq) \rightarrow Zn^{2+}(aq) + 2Ag(s)$$
$$Ag(s) + Zn^{2+}(aq) - \text{no reaction}$$

The higher you go up the series, the more vigorous is the reaction with **water** or **acids**.

Potassium reacts violently with cold water, magnesium reacts *very* slowly with cold water but vigorously with steam. Copper reacts with neither.

See **Electrochemical series**. A series which involves nonmetals.

Recrystallization This is a process for the purification of impure crystalline substances. They are dissolved in a **solvent**, filtered, and then allowed to crystallize. In this way impurities are removed.

Redox A reaction involving both *re*duction and *ox*idation. Any reaction which involves an **oxidation** must also involve a **reduction**.

Examples of redox reactions:

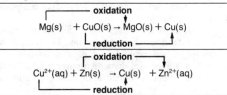

Reducing agent A reducing agent brings about the **reduction** of a substance. In the reaction it is always oxidized. Common reducing agents:

Hydrogen	H_2
Carbon	C
Carbon monoxide	CO
Sulphur dioxide	SO_2
Hydrogen sulphide	H_2S

Reducing agents at work:
$Fe_2O_3(s) + 3CO(g) \rightarrow 2Fe(l) + 3CO_2(g)$
$CuO(s) + H_2(g) \rightarrow Cu(s) + H_2O(g)$
$ZnO(s) + C(s) \rightarrow Zn(s) + CO(g)$

Reduction A substance undergoes reduction if it:

loses **oxygen**	$PbO(s) + C(s) \rightarrow Pb(s) + CO$
gains **hydrogen**	$Cl_2(g) + H_2(g) \rightarrow 2HCl(g)$
gains **electrons**	$Na^+(l) + e^- \rightarrow Na(l)$

See **Oxidation, Reducing agent** and **Redox**.

Refining This involves either the removal of impurities from a substance or the removal of components from a mixture.

(a) *Metals.* When a metal has been extracted from its ores it is often not in a **pure** state. For some uses this might be acceptable, e.g. pig iron contain impurities. Other uses demand a purer product and then the metal is refined until the right purity is achieved.

(b) *Petroleum.* Petroleum is a complex **mixture** of **hydrocarbons**. To make it more useful it is first of all split up into separate *parts*. These *parts* are known as *fractions* and the separation is done by **fractional distillation**. The fractions are also mixtures but are not so complicated. They contain compounds whose **boiling points** are within a certain range.

Conditions vary from refinery to refinery and

depend on what products are required. The major product might be **petrol** or it might be petrochemical feedstock which is then turned into alkenes, etc. for the production of polymers.

See **Cracking** and **Reforming**.

Reforming An important process which is used in petroleum refineries and chemical plants. Molecules are taken and altered into more useful products. There is no change in the size of the molecule. One example is the conversion of unbranched alkanes into branched ones, these giving **petrol** a higher **octane rating**:

octane   2-methylheptane

Another example is the production of **aromatic** compounds from **alkenes**:

$$C_7H_{14} \rightarrow \qquad CH_7H_8 \quad + \quad 3H_2$$
heptene   **methylbenzene** hydrogen

Relative atomic mass (R.A.M.) The relative atomic mass of an element is the mass of an 'average atom' of the element compared with an atom of the $_6^{12}C$ carbon isotope which is given the value of exactly 12. See A_r.

Relative molecular mass (R.M.M.) See M_r.

Residue This is what remains after any chemical process is complete. The material remaining in a filter paper is also called the residue.

Respiration The process by which **energy** is obtained by plants and animals.

In animals, food is eaten and broken down in the body. **Oxygen** is breathed in by taking **air** into the lungs. The oxygen is carried round the body in the blood. Reactions occur between the oxygen and the broken-down food and heat is released. **Carbon dioxide** is also produced and this is breathed out. The waste products from the food are excreted from the body. The energy released is used for heat, movement, growth and all other body functions.

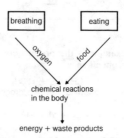

Reversible reactions A reversible reaction is one that can go in either direction, depending on the conditions that exist.

Example:

$$Fe_2O_3(s) + 3H_2(g) \rightleftharpoons 2Fe(s) + 3H_2O(g)$$

Steam can react with hot iron or hydrogen can reduce iron(III) oxide. In both cases, an **equilibrium** will exist with all four substances present unless the products are removed.

Rock salt Impure crystalline deposits of **sodium chloride** are known as rock salt. The solid is purified before use.

Room temperature A temperature of between 15°C and 25°C. It is *not* a fixed temperature, but is a range.

Rubber A natural **polymer**. It is a **hydrocarbon** and its structure is shown here:

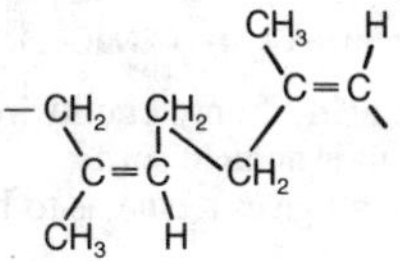

Today most 'rubber' is man-made from butadiene:

$$(CH_2 = CH - CH = CH_2)$$

by **polymerization**. Rubber is elastic and a good insulator. Its uses are widespread from tyres and gloves to waterproofing.

Rust The reddish-brown product of corrosion of

iron which has been exposed to air and water. It is hydrated iron(III) oxide ($Fe_2O_3.xH_2O$). Rusting is of great economic importance. It is most commonly prevented by coating the iron with paint, plastic or another metal. See **Electroplating** and **Galvanizing**.

Salt This is a **compound** which is formed when the **hydrogen** of an **acid** is totally or partially replaced by a **metal**, e.g.:

$$Zn(s) + HCl(aq) \rightarrow ZnCl_2(aq) + H_2(g)$$

When an acid reacts with a base the products are a salt and **water** only:

$$NaOH(aq) + HNO_3(aq) \rightarrow NaNO_3(aq) + H_2O(l)$$

Salts can also be made by direct combination of two elements:

$$2Na(s) + Cl_2(g) \rightarrow 2NaCl(s)$$

With a **dibasic acid**, if only one hydrogen atom is replaced, the result is an **acid salt**.

The name of a salt gives a clue as to how it can be formed, e.g.:
Magnesium sulphate from magnesium and sulphiric acid:

$$Mg(s) + H_2SO_4(aq) \rightarrow MgSO_4(aq) + H_2(g)$$

Saponification This is the process of the **hydrolysis** of an **ester** when *alkaline* conditions are used. The breakdown of natural fats to produce soap is an example of saponification. Here sodium hydroxide solution is used to produce an alcohol and the sodium salt of the **carboxylic acid (soap)**. See **Detergent**.

Saturated compound An **organic** compound containing *only* **single bonds**. Compounds which contain double or triple bonds are said to be **unsaturated**.

Examples:

ethane   butan-l-ol

Saturated solution A **solution** which contains the maximum amount of **solute** *at a given temperature* in the presence of excess **solute** is said to be saturated. The amount of solute needed to form a saturated solution depends on the **temperature**. The only way of putting more solute into a saturated solution is by changing the temperature. See **Supersaturated solution**.

Second (s) The **SI unit** of time.

Semiconductor An electrical **conductor** with unusual **properties**. As the temperature increases or the substance contains greater amounts of impurity the resistance of the material *decreases*. In practice, when crystals of semiconductors are grown, controlled amounts of impurity are added to obtain *exactly* the properties which are required.

Substances which have semiconduction properties can be **elements** or **compounds** but usually involve **metalloids**. Examples include: gallium, germanium, arsenic.

Separation of mixtures Mixtures can be separated if the substances in the mixture have different physical **properties**. Which technique is used depends on the mixture. Techniques include: filtration, fractional distillation, chromatography, decantation.

Shell See **Electronic configuration**.

SI units An international system of **units** based on the metric system of measurement. There are seven basic units (shown below) and the system is used in scientific and technological work throughout the world.

Quantity	Unit	Symbol
• length	metre	m
• mass	kilogram	kg
• time	second	s
• current	ampère	A
• temperature	kelvin	K
• amount of substance	mole	mol
• luminous intensity	candela	cd

Silica gel This is a form of silicon dioxide (SiO_2)

which can absorb up to 40% of its own mass of water. It is used as a **dehydrating agent** in **desiccators** and is also used in packaging to maintain goods in a dry atmosphere. It is possible to add a cobalt(II) salt to the gel in order to indicate the presence of water, e.g.:

red	blue
wet	dry
hydrated	anhydrous
cobalt(II)	cobalt(II)
chloride	chloride
$CoCl_2.2H_2O$	$CoCl_2$

Silicon A **group** (IV) **metalloid**. It is a very abundant **element**, the second most abundant in the earth's crust (27.7%), being a part of the chemical composition of many rocks. It is a **semiconductor** and is the heart of micro-electronic technology — the silicon chip. Silicon is found naturally in the form of metal silicates and silica (silicon dioxide). The mineral quartz is the purest form of silica. Sand is tiny crystals of silica with various impurities.

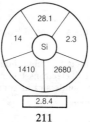

Silver Silver is a **transition metal**. It is used for jewellery and decorative purposes and has been used for coinage. It is often alloyed with **copper** to give strength. It is an excellent conductor of electricity and heat. When exposed to the air it slowly becomes covered in a black film of silver sulphide.

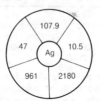

Silver is low in the **reactivity series**. Its oxide is unstable:

$$Ag_2O(s) \xrightarrow{heat} 2Ag(s) + O_2(g)$$

Its **halides** are also unstable. This property is used in photography:

$$2AgBr(s) \xrightarrow{light} 2Ag(s) + Br_2(g)$$

Single bond This is a **covalent bond** which is made up of a shared pair of **electrons**. Examples are found in all **organic** compounds and in compounds formed from nonmetals, e.g.:

212

methane   ammonia   hydrogen chloride

Slaked lime (Ca(OH)₂) This is **calcium hydroxide**. It can be produced by the *slaking* of lime with water:

$$CaO(s) + H_2O(l) \rightarrow Ca(OH)_2(s)$$

A solution of slaked lime in water is known as **limewater**. Slaked lime is used in agriculture and in making mortar. See **Calcium compounds**.

Soap Soap is a cleaning agent which is made by the action of an **alkali**, e.g. sodium hydroxide solution, on naturally occurring **esters**. See **Saponification**.

An example of a soapy detergent molecule.

The **hydrocarbon** end of the molecule is **hydrophobic**. The carboxylate end of the molecule is **hydrophilic**.
See **Detergent**.

Sodium Sodium is a soft, grey **metal** which is in **group** I of the **periodic table**. It is easily cut with a knife revealing a silvery surface which rapidly tarnishes. It is stored under oil because of its reactivity towards air

213

and water. It is very reactive towards water and non-metals e.g.:

$$Na(s) + 2H_2O(l) \rightarrow NaOH(aq) + H_2(g)$$
$$2Na(s) + Cl_2(g) \rightarrow 2NaCl(s)$$

The metal ion gives a yellow/orange flame test.

Sodium is extracted from molten **sodium chloride** by **electrolysis**:

$$2NaCl(l) \rightarrow 2Na(l) + Cl_2(g)$$

The sodium chloride is obtained by mining **rock salt**, as happens in Cheshire. Sodium metal is used as a coolant in fast-breeder nuclear reactors and in the manufacture of the petrol additive tetraethyl lead ($Pb(C_2H_5)_4$) which is used to raise the **octane rating** of fuels. It is also used to extract metals, e.g. titanium:

$$TiCl_4(l) + 4Na(l) \rightarrow Ti(s) + 4NaCl(l)$$

Sodium ions (Na^+) are an important constituent of the fluids in animal tissues. This is why sodium chloride (salt) is added to food during the cooking process.

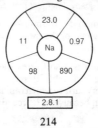

2.8.1

Sodium compounds

Sodium carbonate Na_2CO_3	This compound is made in the **Solvay process** and is an important chemical. One of its many uses is in the manufacture of **glass**. Sodium carbonate **hydrate** shows **efflorescence**: $$Na_2CO_3.10H_2O(s) \rightarrow Na_2CO_3.H_2O(s) + 9H_2O(g)$$ Unlike most carbonates, it is soluble in water. It is not decomposed by heat, its aqueous solution is *alkaline*.
Sodium chloride $NaCl$	Known as *common salt* this compound is obtained from **rock salt** and is used to make **sodium hydroxide, sodium metal** and **chlorine**. It is used for flavouring and preserving foods. See **Solvay process**.
Sodium hydrogen-carbonate $NaHCO_3$	This is used in **baking powder**. It is decomposed by the action of heat or **acids**. $$\overset{\text{heat}}{2NaHCO_3 \rightarrow Na_2CO_3(s) +} CO_2(g) + H_2O$$ $$NaHCO_3(s) + HCl(aq) \rightarrow NaCl(aq) + H_2O(l) + CO_2(g)$$ Sodium hydrogencarbonate is also used in **fire extinguishers** and in **anti-indigestion powders**.

Sodium hydroxide NaOH	This is produced by the electrolysis of **brine** and has wide uses in industry, e.g. making soap and paper. It is a **caustic alkali**. Solutions have pH > 10.
Sodium nitrate $NaNO_3$	This is used as a fertilizer and in the preservation of meat.
Sodium sulphate Na_2SO_4	The hydrated salt ($Na_2SO_4.10H_2O$) is known as Glauber's Salt. The sulphate is used in paper manufacture.
Sodium thiosulphate $Na_2S_2O_3$	This compound is used in the photographic process. It is used to *fix* the negative and is often known as *hypo*. It reacts with unreacted silver bromide and therefore makes sure that no further reaction with light occurs.

Softening See the **Hardness of water**.

Solder An alloy of tin and lead in various proportions depending upon the use, e.g. plumber's solder, soft solder for electrical connections, etc.

Solid A solid is a substance whose **atoms** or **molecules** are fixed in positions and do not have the freedom of movement found in a **liquid** or a **gas**. Atoms and molecules are held in a **lattice** by **bonds**. It is only when these bonds are broken that the atoms and molecules are able to move and the solid *melts*.

216

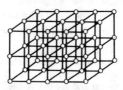

As **heat energy** is put into the solid lattice the atoms or molecules acquire enough energy to break the bonds. Although the atoms or molecules within the lattice do not move from place to place they do vibrate.

Soluble A substance is soluble if it will **dissolve** in a **solvent**. The extent to which any **solute** dissolves depends on the solvent and the temperature. At any given temperature there is a maximum amount of solute which can dissolve in a fixed volume of solvent. This produces a **saturated solution**. See **Solubility**.

Solubility The solubility of a **solute** in a **solvent** is the extent to which it *dissolves*. An **ionic** substance will have a higher solubility in a **polar** solvent than a covalently bonded solute, e.g. **copper (II) sulphate** is much more soluble in **water** than is **methane**. The **units** used are **moles** of solute in 100 g of solvent at a *stated temperature*, although other units, e.g. mole/dm³, g/100 g, are also used.

Solute A solute is the name given to substances which **dissolve** into a **solvent** to produce a **solution**, e.g. when copper(II) sulphate dissolves in water to

217

produce a solution, copper(II) sulphate is a solute:

$$CuSO_4(s) + H_2O(l) = CuSO_4(aq)$$

Solution When a solid, liquid or gas **dissolves** in a **solvent** to form a solution the particles of the **solute** (i.e. its atoms or molecules) are spread throughout the body of the solvent and are not visible.

Solvay process This process takes cheap raw materials (**brine** and **calcium carbonate** — **limestone**) and turns them into the valuable alkali, **sodium carbonate**:

$$2NaCl(aq) + CaCO_3(s) \rightarrow$$
$$CaCl_2(aq) + Na_2CO_3(aq)$$

Because both sodium carbonate and calcium chloride are soluble in water, the reaction shown cannot be carried out *directly*. It does work, however, by a series of steps:

(a) Limestone is heated:

$$CaCO_3(s) \rightarrow CaO(s) + CO_2(g)$$

(b) Ammonia is dissolved in brine ($NaCl + NH_3$).
(c) Carbon dioxide (produced from the limestone) is passed into the brine/ammonia solution. The following reaction occurs:

$$NaCl(aq) + CO_2(g) + NH_3(g) + H_2O(l) \rightarrow$$
$$NaHCO_3(s) + NH_4Cl(aq)$$

Sodium hydrogencarbonate is produced because of its low solubility. It can be removed from the mixture.

218

(d) Sodium carbonate is produced by heating the hydrogencarbonate:

$$2NaHCO_3(s) \rightarrow Na_2CO_3(s) + H_2O(l) + CO_2(g)$$

(e) In the final stage the ammonia is recovered from the ammonium chloride:

$$2NH_4Cl(aq) + CaO(s) \rightarrow$$
$$CaCl_2(aq) + 2NH_3(g) + H_2O(l)$$

The ammonia and carbon dioxide produced in these last two stages are recycled and used again. The only by-product of the reaction is calcium chloride ($CaCl_2$).

Solvent This is a **liquid** in which a **solute dissolves** to form a **solution**. **Water** is the most common solvent. Solvents may be polar, e.g. water, or nonpolar, e.g. **ether**. Polar solvents dissolve ionic or polar solutes, e.g. **salts**. Nonpolar solvents dissolve **covalent** molecules, e.g. **hydrocarbons**.

Spectator ions These are ions which play no part in a reaction. They are only *spectators*. In the example shown below the spectator ions are underlined.

$$\underline{Na^+(aq)} + Cl^-(aq) + Ag^+(aq) + \underline{NO_3^-(aq)} \rightarrow$$
$$\underline{Na^+(aq)} + \underline{NO_3^-(aq)} + AgCl(s)$$

The reaction is: $\boxed{Ag^+(aq) + Cl^-(aq) \rightarrow AgCl(s)}$

Stalactites and stalagmites These are growths

which are found in caves. When water containing **calcium salts** falls into the cave there is a gradual build up of **calcium carbonate** from the roof (forming stalactites) and onto the floor (forming stalagmites) as the solution containing the calcium salts decomposes.

$$Ca(HCO_3)_2(aq) \rightarrow CaCO_3(s) + H_2O(l) + CO_2(g)$$

Starch ($(C_6H_{10}O_5)_n$) **A carbohydrate polymer**. It is made up of **glucose monomers** and can be broken down by the action of **enzymes** or dilute **acid**. Glucose or **maltose** is formed. Starch is insoluble in water and is a white tasteless powder. It is found in most plants and is used by plants as a food. Animals also use it as food when they eat plants, e.g. potatoes, rice, flour.

In the presence of starch, **iodine** solution turns blue. This can be used as a test for iodine or starch. See **Polysaccharide**.

States of matter Substances can exist in three states of matter. These are **solid, liquid** and **gas**. Substances are changed from one state into another by altering the **temperature**.

State symbols These are the letters placed next to the formula of a substance, usually in brackets, in a chemical equation to denote the state of matter of that substance in the reaction. They are: (s) = solid, (l) = liquid, (g) = gas, (aq) = aqueous solution. E.g.:

$$2Na(s) + 2H_2O(l) \rightarrow 2NaOH(aq) + H_2(g)$$

Steel Steels are **alloys** which contain **iron** as the main constituent. Depending on which other **elements** are present, substances of varying properties can be formed. The two best known steels are *mild steel* which is used for car bodies and household goods such as cookers, freezers, etc. and *stainless steel* which is used in industry and in cooking utensils. Mild steel rusts easily and so has to be protected by **galvanizing**, enamelling or painting. Stainless steel is not corroded by oxygen and so is very useful.

The iron made in the blast furnace contains impurities which make it brittle, e.g. carbon. Iron is made into steel by blowing oxygen through the molten iron and thus oxidizing the impurities. Then, the correct amounts of other elements can be added to produce the steel required, e.g. stainless steel.

STP This stands for *S*tandard *T*emperature and *P*ressure. The conditions it refers to are one **atmosphere** pressure and a **temperature** of 0°C (273 K). When comparing gas volumes it is useful to have *standard conditions* to refer to.

Strengths of acids and bases **Acids** and **bases** are termed *strong* if they are fully dissociated in solution. Good examples are acids such as hydrochloric, nitric and sulphuric and bases such as sodium or potassium hydroxide.

$$HCl(aq) \rightleftharpoons H^+(aq) + Cl^-(aq)$$
pH range 1—2

$$NaOH(aq) \rightleftharpoons Na^+(aq) + OH^-(aq)$$
pH range 13—14

Weak acids and bases do not fully dissociate in solution. Here the equilibria lie to the *left-hand* side. Good examples are found in the organic acids, e.g. **citric**, tartaric, **ethanoic** and malic; and in bases such as ammonia.

$$CH_3COOH(aq) \rightleftharpoons CH_3COO^-(aq) + H^+(aq)$$
pH range 3—6
$$NH_3(aq) + H_2O(l) \rightleftharpoons NH_4^+(aq) + OH^-(aq)$$
pH range 8—11

Strong acid and base See **Strengths of acids and bases.**

Structural formula A structural formula is one that shows the **bond** between atoms and the position of the atoms with respect to each other. The **molecular formula** shows the number of atoms in the molecule.

ethane
C_2H_6

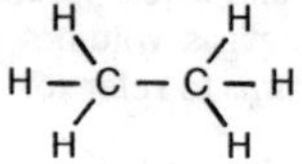

ethanol
C_2H_5OH

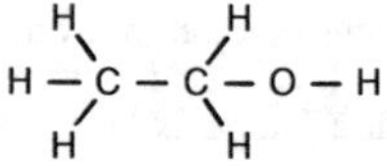

molecular formula   structural formula

Subatomic particles These are particles which are

222

found in the atom. The important ones are the **proton**, **neutron** and **electron**.

Sublime If a **solid** changes directly into a **gas** when it is heated, or if the vapour changes directly into a solid on cooling, it is said to sublime. Common substances which sublime are:

Carbon dioxide CO_2 Iron(II) chloride $FeCl_2$

Substitution reaction This is a reaction where some atoms or groups in a molecule are replaced by others. It is a common reaction of **organic** compounds.

Example:

$CH_4(g) + Cl_2(g) \rightarrow CH_3Cl(l) + HCl(g)$
A C—H bond is replaced by a C—Cl bond.

Sucrose ($C_{12}H_{22}O_{11}$) A **disaccharide** which is made up of a **fructose** unit and a **glucose** unit. It is the white crystalline **sugar** that is used in the home. It is obtained from sugar beet and cane sugar.

Sulphates These compounds contain the SO_4^{2-} **ion** (valency = 2). They are widespread in nature, e.g. gypsum — $CaSO_4$. They can be produced in the laboratory by the action of **sulphuric acid** on metals or oxides and hydroxides, e.g.:

$Zn(s) + H_2SO_4(aq) \rightarrow ZnSO_4(aq) + H_2(g)$
$CuO(s) + H_2SO_4(aq) \rightarrow CuSO_4(aq) + H_2O(l)$

The test for a sulphate is to add a solution to an acidified (HCl) solution of barium chloride. If a sulphate is present a white **precipitate** is formed:

$$BaCl_2(aq) + SO_4^{2-}(aq) \rightarrow BaSO_4(s) + 2Cl^-(aq)$$

This precipitate is insoluble in dilute acid.

Sulphides Compounds of sulphur with another **element**. They are produced by reaction with **sulphur** or **hydrogen sulphide**:

$$Fe(s) + S(s) \rightarrow FeS(s)$$
$$CuSO_4(aq) + H_2S(g) \rightarrow CuS(s) + H_2SO_4(aq)$$

Sulphites These compounds contain the SO_3^{2-} **ion** and can be thought of as **salts** of **sulphurous acid** H_2SO_3. **Sulphur(IV) oxide** is prepared in the laboratory by the action of **acid** on a sulphite.

Sugar This is the common general term for those sweet compounds which chemically are **monosaccharides** and **disaccharides**. Examples: **glucose, sucrose, fructose, maltose**. See **Carbohydrate**.

Sulphur (S_8) Sulphur is a yellow nonmetallic **element** which exists as two allotropic forms. Rhombic sulphur is the stable form below 96°C and monoclinic sulphur is stable above that temperature. The figures in the chart refer to the monoclinic **allotrope**.

Sulphur is in **group** (VI) of the **periodic table** and is reactive towards metals and oxygen. It is found uncombined in nature as well as occurring as metal sulphides, e.g. galena (PbS) and pyrite (FeS).

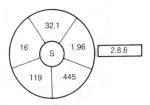

Sulphur-containing compounds are also found in **petroleum** and **natural gas**. Sulphur is an important constituent of some drugs, e.g. sulphonamides but its major use is in the manufacture of **sulphuric acid** in the **contact process**. See **Vulcanizing**.

Sulphur compounds

Sulphur(IV) oxide SO_2	This is commonly called sulphur dioxide and is a colourless gas with a sharp odour. It is produced by burning **sulphur** or sulphur compounds in **air** or **oxygen**. It dissolves in water to form **sulphuric acid** — a weak acid. It is useful as a sterilizing agent and is often added to soft drinks. It is a **reducing agent**. This property is used in the chemical test for the gas. Sulphur(IV) oxide will *reduce* dichromate(VI) ion (orange) to chromium(III) ion (green). Sulphur dioxide is a major cause of air pollution and acid rain, being

| Sulphur(VI) oxide SO_3 | This is a **volatile** solid with a sharp odour. It is produced in the **contact process** by the combination of **sulphur dioxide** and **oxygen** over a **catalyst** (Vanadium pentoxide). |
| | The oxide is acidic and when it is added to water, sulphuric acid is formed. |

produced by the combustion of fossil fuels.

$$H_2O(l) + SO_3(s) \rightarrow H_2SO_4(aq)$$

Hence it is the anhydride of sulphuric acid.

Sulphuric acid (H_2SO_4) One of the most important chemicals produced. It is a colourless, oily liquid which is a **strong acid** and a vigorous **oxidizing agent**. It is made in the **contact process** and the annual production in the UK is about four million **tonnes**. The chart opposite shows the main uses of the acid and the approximate amounts of the acid which are used.

Sulphuric acid reacts chemically in several ways. *As an acid:* dilute sulphuric acid reacts with **metals, bases** and **carbonates** to form sulphates:

$$Mg(s) + H_2SO_4(aq) \rightarrow MgSO_4(aq) + H_2(g)$$
$$CuO(s) + H_2SO_4(aq) \rightarrow CuSO_4(aq) + H_2O(l)$$
$$ZnCO_3(s) + H_2SO_4(aq) \rightarrow$$
$$ZnSO_4(aq) + H_2O(l) + CO_2(g)$$

Concentrated sulphuric acid reacts with chlorides

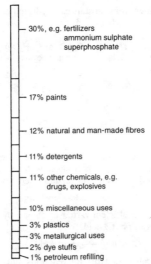

- 30%, e.g. fertilizers
  ammonium sulphate
  superphosphate

- 17% paints

- 12% natural and man-made fibres

- 11% detergents

- 11% other chemicals, e.g.
  drugs, explosives

- 10% miscellaneous uses

- 3% plastics
- 3% metallurgical uses
- 2% dye stuffs
- 1% petroleum refilling

and nitrates to form hydrogen chloride and nitric acid respectively:

$$H_2SO_4(l) + NaCl(s) \rightarrow NaHSO_4(s) + HCl(g)$$
$$H_2SO_4(l) + NaNO_3(s) \rightarrow NaHSO_4(s) + HNO_3(g)$$

As a dehydrating agent: the concentrated acid is **hygroscopic**. Concentrated sulphuric acid is sometimes used to dry gases. It can also be used to

227

remove water from substances. Hydrated copper(II) sulphate can be dehydrated:

$$CuSO_4.5H_2O(s) \rightarrow CuSO_4(s) + 5H_2O(l)$$
blue crystals   white powder

methanoic   carbon
acid   monoxide

If the concentrated acid is added to **sucrose**, the sugar turns black. This is because water is removed from the **carbohydrate** and **carbon** is left behind:

$$C_6H_{12}O_6(s) \rightarrow 6C(s) + 6H_2O(g)$$

Similarly ethanol is converted to ethene by heating with excess sulphuric acid:

$$C_2H_5OH(l) \rightarrow C_2H_4(g) + H_2O(l)$$

These are called 'dehydration reactions' since a water molecule has been eliminated from the reactant molecule.

The reaction between concentrated sulphuric acid and water is very **exothermic**. It is important always to add the acid to water and *not* the other way round. *As an oxidizing agent:* although copper cannot displace hydrogen from acids, the metal can be oxidized by concentrated sulphuric acid:

$$Cu(s) + 2H_2SO_4(l) \rightarrow CuSO_4(s) + SO_2(g) + 2H_2O(l)$$

anhydrous
copper(II)
sulphate

Sulphurous acid (H_2SO_3) This acid only exists in aqueous solution. It is formed by passing **sulphur dioxide** into water.

It is a **weak acid** and a **reducing agent**. When heated, sulphur dioxide and steam are produced, leaving no residue. Salts of the acid are called **sulphites.**

Supersaturated solution This is a solution which contains a higher **concentration** of **solute** than a **saturated solution**. It is usually produced by cooling the saturated solution. If the solution is disturbed, e.g. by a mechanical shock or by dust falling into it or by a crystal of the solute being added to it, the excess solute will usually crystallize out.

Suspension When a **solid** is added to a **liquid** and the solid neither dissolves in the liquid nor sinks to the bottom of the vessel, the mixture is referred to as a suspension because the solid is *suspended* in the liquid.

solid particles
suspended in
the liquid

Symbol Symbols are often used in chemistry to

represent the names of substances, e.g. **elements**.

$$Fe - Iron$$
$$S - Sulphur$$
$$O - Oxygen$$

Symbols are also used in representing units.

Synthesis This term means the production of something from smaller or simpler parts, e.g. the synthesis of **ammonia** from hydrogen and nitrogen.

Synthesis gas This is a mixture of **carbon monoxide** and **hydrogen** gas produced by the steam **reforming** of **natural gas**. It is used to synthesize other compounds.

Synthetic Artifical, e.g. artificial fibres such as **nylon**, etc.

Temperature The temperature of a substance is the measure of its **kinetic energy**. The temperature scale used in everyday life is the *Celsius* (or *Centigrade*) scale. In scientific work, the **Kelvin** scale is usually used.

Tempering This is when a hot **metal** is allowed to cool slowly in air. By doing this, the metal becomes tough and springy.

Terylene This is a **polyester** which is widely used in the clothing and household furnishing industries. It is often mixed with natural fibres, e.g. Terylene/wool mixtures. Its **monomers** are a dialcohol and a diacid. The formula of the polymer is:

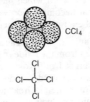

| from the | from the |
| diacid | dialcohol |

Tetrachloromethane Tetrachloromethane is a dense, colourless liquid which is **immiscible** with water. It is a good solvent for non-**polar** compounds.

It can be made from methane by a **substitution reaction** involving **chlorine**:

$$4Cl_2(g) + CH_4(g) \rightarrow CCl_4(l) + 4HCl(g)$$

The other substituted methane products are also formed: (CH_3Cl, CH_2Cl_2, $CHCl_3$) and the mixture must be separated to obtain **pure** products.

Tetrahedral In a **carbon compound** such as **methane** where there are four **single bonds** leading to four atoms, the atoms are arranged tetraherally around the carbon atom. The bonds point to the corners of the imaginary cube that the carbon atom is in the centre of. In this way, the atoms are as far away from each other as possible. The angle between the bonds is 109° 28′ — tetrahedral angle.

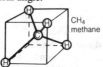

CH₄
methane

Thermal This is a word which means 'to do with heat', e.g. thermal decomposition — decomposition by heat. Thermal **energy** — heat energy.

Thermit reaction In this reaction aluminium removes the oxyen from the oxide of a metal below it in the **reactivity series**. Iron(II) oxide is usually used in the laboratory and was used in industry to provide a small source of molten metal. Great heat is liberated in the reaction and the products are iron and aluminium oxide.

$$Fe_2O_3(s) + 2Al(s) \rightarrow Al_2O_3(s) + 2Fe(l)$$

In the reaction in the laboratory a 'fuse' of magnesium is usually used and barium peroxide is used to start the reaction off. Once started, the reaction is very **exothermic**.

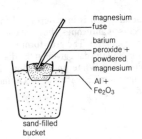

magnesium
fuse

barium
peroxide +
powdered
magnesium

Al +
Fe_2O_3

sand-filled
bucket

This reaction can be used for welding cracks in iron.

Thermometer Thermometers measure **temperature** — i.e. how hot something is. Although they come in different forms, the most common are the 'liquid in glass' thermometers.

The liquid — usually **mercury** — is held in a bulb. As the temperature rises, the liquid expands in the bulb and rises up the tube. The tube is marked at intervals with the temperature and this can be read off.

Thermoplastic A thermoplastic **polymer** is one which softens when it is heated and can be moulded and re-moulded into new shapes. Examples: **nylon, poly(chloroethene)**.

Thermosetting A thermosetting **polymer** is one which cannot be softened again once it has been heated. Decomposition occurs if it is heated again. Examples: bakelite, formica.

Tin A **metal** which is in group IV of the **periodic table**. It occurs as **allotropes**, the most common being grey tin and white tin. The grey form is stable at low temperatures and the data given in the diagram refers to the white form. White tin is a shiny metal which shows normal metallic reactions. It is found naturally as the oxide SnO_2.

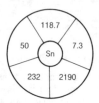

Tin is used to make tin plate (**steel** with a very thin layer of tin upon it), **bronze**, solder and **alloys** for bearings. It is also used in the production of **glass**. Glass is floated onto the surface of molten tin. In this way large areas of glass without defects are formed. Tinplate is used to make cans and metal boxes in which food is stored.

Titanium A **transition metal** which is common in the earth's crust (0.5%). It forms **alloys** which are very strong while having a low **density**. It is resistant to **corrosion** up to a high temperature. Titanium alloys are used in the aircraft industry. TiO_2 is used as a pigment. See diagram opposite.

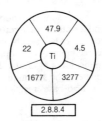

2.8.8.4

Titration A titration involves the reaction of two **solutions** and is used in analysis to discover the **concentration** of one of the solutions. An accurately

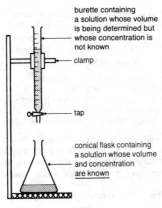

burette containing
a solution whose volume
is being determined but
whose concentration is
not known

clamp

tap

conical flask containing
a solution whose volume
and concentration
<u>are known</u>

measured volume of one solution is normally placed in a conical flask and the other in a **burette**. The solution from the burette is added slowly until the other solution has been completely used up. This is shown by the use of an **indicator** or by means of a **pH** meter or a conductivity meter. The technique is most commonly used for acid-base reactions. See **Volumetric analysis**.

Tonne This is a **unit** of mass. One tonne equals 1000 kilograms. It is 2205 lb and so is 35 lb lighter than the imperial ton (2240 lb). The tonne (sometimes called the *metric ton*) is now widely used.

Transition metals These are the **elements** found in the central section of the **periodic table**. There are three series of them but the most important are those elements from scandium to zinc. Transition metals have similar **properties**. They

- produce coloured **compounds**
- have variable **valency**
- are high **melting** and **boiling point** solids
- are useful as **catalysts**
- have **cations** which are useful as catalysts.

They tend to be important elements for use as metals, e.g. iron, copper, silver, gold, mercury, zinc, platinum, or for use in alloys, e.g. titanium, vanadium, chromium, manganese, cobalt, nickel, tungsten.

Transition temperature The **temperature** above

which one **allotrope** is stable and below which another is stable. Some examples are shown here.

rhombic sulphur stable below	96°C	monoclinic sulphur stable above
grey tin stable below	13°C	white tin stable above

Trichloromethane This molecule is also known as *chloroform* and has been used in anaesthetic work.

Trichloromethane is a substituted alkane. It is a colourless, volatile liquid which boils at 62°C. See **Tetrachloromethane**.

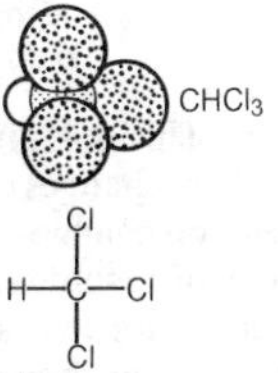

CHCl₃

H — C — Cl structure with Cl atoms

Triple bond A triple bond contains three shared pairs of **electrons** and is found in **alkynes** and nitrogen-containing compounds.

Examples:

ethyne H — C ≡ C — H
nitrogen N ≡ N

Ultraviolet radiation (UV) This is invisible radiation

which has a slightly higher frequency (energy) than violet light. It is produced in large amounts by stars, e.g. the sun. The radiation in large amounts is harmful to humans — skin cancers can result. Not much of the UV radiation from the sun reaches the earth. Much of it is filtered out by the **ozone** layer in the upper **atmosphere**.

Units   These are quantities or measurements which are used as standards to measure other things. See **SI units**.

Measurement	Unit
distance	metre
time	second
mass	kilogramme

Universal indicator   This is a mixture of **indicators**. Because it is a mixture it changes colour several times as the pH of the solution changes. It is possible to tell the approximate **pH** of a solution by adding a few drops of universal indicator to it and reading the pH off a chart such as the one shown below. Universal indicator paper strips are also available:

Colour	← red orange yellow green blue purple →
pH	0 1 2 3 4 5 6 7 8 9 10 11 12 13 14

Unsaturated compounds   Carbon compounds which possess **double** or **triple bonds** between two carbon atoms are said to be unsaturated. Some examples are:

238

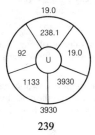

$$\underset{\text{alkenes}}{\overset{\diagdown}{\underset{\diagup}{C}}=\overset{\diagup}{\underset{\diagdown}{C}}} \qquad \underset{\text{alkynes}}{-C \equiv C-}$$

They contain more **electrons** in their bond than a normal **single bond**. Because of this they are reactive. They tend to react through **addition reactions** forming new bonds with the electrons they possess, e.g. **alkenes** become **alkanes**.

$$\underset{\text{ethene (unsaturated)}}{\overset{H}{\underset{H}{\diagdown}}C=C\overset{H}{\underset{H}{\diagup}}} \quad \xrightarrow{H_2} \quad \underset{\text{ethane (saturated)}}{H-\overset{\overset{\displaystyle H}{|}}{\underset{\underset{\displaystyle H}{|}}{C}}-\overset{\overset{\displaystyle H}{|}}{\underset{\underset{\displaystyle H}{|}}{C}}-H}$$

Uranium A metallic element which has three naturally occurring **isotopes** (masses 234 — trace, 235 — 0.7%, 238 — 99.3%). They are all radioactive and eventually decay to give stable isotopes of **lead**.

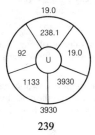

239

Uranium is used as a fuel in nuclear power stations. In some types the oxide U_3O_8 is used. In others uranium containing a higher proportion of U-235 is used (enriched uranium). In the fast-breeder reactor uranium 238 is turned into **plutonium** 239 which can then be used as a fuel.

Vacuum A vacuum is where there are no **atoms** or **molecules** present. It is impossible to obtain a *perfect* vacuum but we talk of a *partial* vacuum when the **pressure** is extremely low. The pressure is low in space and this is usually thought of as a vacuum.

Valency Valency can be simply thought of as *combining power* i.e. the usual number of bonds which an atom forms when making compounds. More precisely, the valency of an **element** is the number of **electrons** that it needs to form a **compound** or radical. The electrons can be *given to* another element; they can be *taken from* an element; they can be *shared*.

Some elements always have the same valency, e.g. hydrogen = 1, oxygen = 2 (except in peroxides), sodium = 1, magnesium = 2.

Examples:
Sodium has a valency of one. Sodium gives one electron away when it forms the sodium **ion** Na^+, e.g. NaCl.

Oxygen has a valency of two. Oxygen accepts two electrons when it forms the oxide ion O^{2-} or it forms covalent compounds, e.g. SO_2, CO_2, N_2O.

240

Hydrogen and bromine have valency one. They both give one electron to make the H−Br **covalent bond** when they form hydrogen bromide.

Transition elements however have more than one valency: Iron = 2, 3; cobalt = 2, 3; copper = 1, 2. The valencies of some common elements and ions are shown in the following tables.

Positive values

+ 1		+ 2		+ 3	
Lithium	Li+	Calcium	Ca2+	Aluminium	Al3+
Sodium	Na+	Magnesium	Mg2+	Iron(III)	Fe3+
Potassium	K+	Zinc	Zn2+		
Silver	Ag+	Iron(II)	Fe2+		
Ammonium	NH4+	Lead	Pb2+		
Hydrogen	H+	Copper(II)	Cu2+		
Copper(I)	Cu+				

Negative values

− 1		− 2		− 3	
Fluoride	F−	Sulphide	S2−	Phosphate	PO43−
Chloride	Cl−	Oxide	O2−		
Bromide	Br−	Carbonate	CO32−		
Iodide	I−	Sulphate	SO42−		
Hydroxide	OH−				
Nitrate	NO3−				

Vanadium Vanadium is a **transition metal**. Its chief use is in the production of vanadium-steel **alloys** which are valuable because of their high tensile strength and hardness. They often include **chromium** or **manganese** in addition to vanadium.

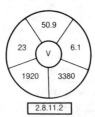

2.8.11.2

Vanadium(v) oxide V_2O_5 is the **catalyst** used in the **contact process.**

Van der Waals' forces These are weak forces between **atoms**. They are caused by the movement of **electrons** within the atoms. Elements which only have these forces to hold them together, e.g. the **inert gases**, have very *low* **melting** and **boiling points.**

Vapour A vapour is **atoms** or **molecules** in the gaseous state *but* below their critical temperature that is the temperature above which liquid cannot exist.

For example, when water evaporates from a saucer when it is left in the house, *water vapour* is formed and not steam.

Vinegar This is a solution which is made by the action of bacteria on wine or cider. It contains about 4% **ethanoic acid**. It is used widely in the food industry for preserving foods.

Vitamin Vitamins are chemicals which are important to the proper working of the body. They tend

242

to be complex **organic** molecules which cannot be made in the body but which must be eaten, as in dairy products (vitamin A) or fruit (vitamin C).

Volatile A substance which is volatile is easily turned into a **vapour**. Such substances either have **boiling points** which are near **room temperature**, e.g. **ether** or propanone; or they are solids which **sublime**, e.g. **carbon dioxide**. Liquids which are volatile *and* flammable (see **flame**) are very dangerous because of the risk of **explosions**. Care has to be taken with their storage and handling. **Petrol** is such a liquid.

Volume The volume of a substance is the amount of space that it occupies. Solids and liquids have fixed volumes but a gas will have the same volume as the container it occupies. The larger the container is, the *lower* is the **pressure**. The volume of a gas can easily be changed by compressing it but it is much more difficult to change the volumes of solids and liquids. *Hydraulic* brakes in a car depend on this property of a liquid. Volumes are measured in cubic centimetres (**cm^3**).

Volumetric analysis This is a method of **quantitative** analysis which uses accurately measured volumes of solutions. See **Titration**, **Burette** and **Pipette**.

Vulcanization In this process **sulphur** is added to **rubber** to make it harder.

Washing soda This name is given to **hydrated**

sodium carbonate ($Na_2CO_3.10H_2O$). The name comes from the use of the salt to soften water which was to be used for washing.

Water (H_2O) Water is an oxide of **hydrogen**:

$$2H_2(g) + O_2(g) \rightarrow 2H_2O(l)$$

It is one of the most common compounds on earth. It is the best known **solvent** and is needed by all living things.

Water is a colourless liquid and some of its more important properties are shown here:

> Freezing point 0°C
> Boiling point 100°C
> Density 1.0 g/cm³
> (water has a maximum density at 4°C)

Unusually, water expands on solidification. This accounts for ice floating and water pipes bursting. It does not conduct electricity although it can be electrolysed if small amounts of acid (H_2SO_4) or alkali (NaOH) are added. The products are hydrogen and oxygen:

$$2H_2O(l) \rightarrow 2H_2(g) + O_2(g)$$

Two tests for the presence of water are by changing the colour of **anhydrous salts**:

$$CuSO_4(s) + 5H_2O \rightarrow CuSO_4.5H_2O(s)$$
white   blue

$$CoCl_2(s) + 6H_2O \rightarrow CoCl_2.6H_2O(s)$$
blue   red

These tests show that water is present and not that the water is **pure**. To show purity, the **boiling point** of the liquid could be taken.

Water is found in the **atmosphere**, in lakes, rivers, glaciers and the oceans. It is found in rocks and in living creatures. Water is continually moving from place to place on the earth — see the **water cycle**.

The most important chemical property of water is its use as a **solvent**. Water has **polar bonds** and so can dissolve ionic solids such as sodium chloride (NaCl) as well as polar solids such as **glucose** ($C_6H_{12}O_6$).

The water which we drink is never pure. It always contains small amounts of gas (e.g. oxygen and carbon dioxide) and depending on the source of the water, solids are dissolved in it too, some of which make the water hard. See **Hardness of water**.

Water cycle There is a continual movement of water around the earth, both in the oceans and in the **air**. Water falls to earth as rain, snow, hail, sleet, and freezes out of the air as frost and ice. It falls onto the oceans and these act as a vast source of water. It falls on land and here it enters the earth and eventually flows into lakes and rivers and then flows into the oceans. From these large areas of water, evaporation occurs and water re-enters the **atmosphere**. In this way the cycle continues. Plants and animals take water out of the ground for their own use and it can re-enter the atmosphere (*transpiration, excretion* and **respiration**) or the earth. Large quantities of water are

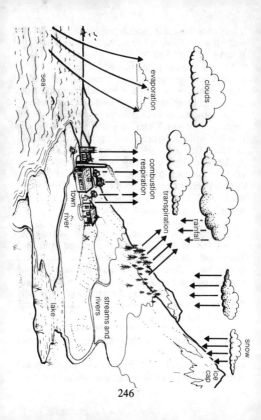

stored as glaciers and the polar ice caps.

Water of crystallization This is water which is chemically bonded within **crystals**.

Examples are:

$CuSO_4.5H_2O$ $Na_2CO_3.10H_2O$ $CaSO_4.2H_2O$

Water can be removed by heating, leaving the **anhydrous** salt. This can happen because the chemical bonds are weak. Some **salts** which include water of crystallization lose some of the chemically bounded water simply on exposure to the air. This is called **efflorescence**.

Weak acids and bases See **Strengths of acids and bases**.

X-rays This is *electromagnetic radiation* which has high **energy**. It is produced by firing **electrons** at **metals**. It is a very penetrating radiation and will easily pass through flesh but is stopped by bone and other dense substances such as metals.

Xenon Xenon is one of the few noble gases which have been found to form compounds. Even so only the two most reactive elements, flourine and oxygen, have succeeded, e.g. XeF_2, XeF_4, XeF_6, XeO_3. The element is used to fill light bulbs and fluorescent tubes.

Yeast Yeasts are microscopic organisms which are very useful to man. They are used in baking and brewing. In brewing, they are used to convert **sugars** into **ethanol**:

$$C_6H_{12}O_6(aq) \xrightarrow{\text{yeast}} 2C_2H_5OH(aq) + 2CO_2(g)$$

glucose	**ethanol**	**carbon dioxide**

The results are alcoholic drinks such as beer and wine. The carbon dioxide is either collected and sold or it is allowed to escape into the **atmosphere**.

In baking, it is the carbon dioxide which is useful. The dough is mixed and then the sugar and yeast react in the dough to produce the gas. The gas makes the dough expand (rise) and the mixture becomes much lighter. Yeasts contain **enzymes** and it is these substances which act on the sugars. See **Zymase**.

Yield of a reaction Many chemical reactions do not produce as much product as would be expected from looking at the chemical **equations**. It is usual to express the amount of product as a percentage of what it is theoretically possible to produce:

$$\text{Yield of reaction} = \frac{\text{amount of product produced}}{\text{maximum amount of product that it is possible to produce}}$$

Example:

$$CuO(s) + H_2(g) \rightarrow Cu(s) + H_2O$$

copper(II)   hydrogen   copper   water
oxide

80 g $\rightarrow$ 64 g (maximum yield)

If 80 g of oxide actually produce 56 g of copper metal the percentage yield of the reaction is $(\frac{56}{64} \times 100)\% =$ 87.5%.

Z This is the symbol which is given to represent the **atomic number** of an **element**, i.e. the number of **protons** in the **atom**.

Zinc Zinc is a **transition metal** which is found in the *first* transition series in the periodic table.

It is a dense grey metal which is reactive towards **acids** but which does not react with cold water. The metal is used in the **alloy, brass,** and also in the protection of **steel** by **galvanizing**. Zinc is extracted from the ore *sphalerite* (zinc blende) (ZnS).

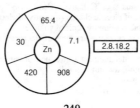

249

Zinc compounds

Zinc oxide ZnO	This has a use in medicine — zinc oxide cream. It is used as a protection against skin irritations, e.g. nappy rash. It is also used in paints. Zinc oxide and hydroxide are **amphoteric**.

Zymase This is the **enzyme** present in **yeasts** which is responsible for the formation of **ethanol** and carbon dioxide from **sugars**.

APPENDIX A

A list of some useful common abbreviations and symbols you may encounter in scientific literature.

A	mass number; also Ampère — unit of electric current
aq	state symbol for aqueous solution usually as (aq)
A_r	relative atomic mass
atm	atmosphere — a unit of pressure
α	alpha Greek letter
ß	beta Greek letter
b.p.	boiling point
C	Celsius as in °C degree Celsius; also Coulomb — unit of electric charge (quantity of electricity)
cm^3	cubic centimetre, unit of volume
DC	direct current — the type of electricity produced from a battery
dm^3	cubic decimetre ≡ 1 litre, unit of volume
E	symbol for emf of a cell
e or e^-	electron
emf	electromotive force

g	gram — unit of mass; state symbol for gas usually as (g); also acceleration due to gravity
H	enthalpy (ΔH = enthalpy change)
I	electric current
J	Joule — unit of energy
k	prefix meaning 'one thousand times' i.e. kg = 1000 g
K	Kelvin — unit of temperature, $1K \equiv 1°C$
l	state symbol for liquid usually as (l)
m	mass; also metre — unit of length
M	molar — unit of concentration (molarity) e.g. 2M
m^3	cubic metre — unit of volume
mol.	mole — unit of amount of substance
ml	milliletre, $\frac{1}{1000}$ of 1 litre $\equiv 1 \text{ cm}^3$
m.p.	melting point
M_r	relative molecular mass
n	neutron
N	Newton — unit of force
NTP	Normal temperature and pressure
p	proton; also pressure
P_a	Pascal — unit of pressure
p.d.	potential difference
pH	relates to a scale of acidity, e.g. pH = 1 very strongly acidic
Q	electric charge, quantity of electricity
s	state symbol for solid usually as (s); also second — unit of time
STP	standard temperature and pressure
t	time
T	temperature
$T_{\frac{1}{2}}$ or $t_{\frac{1}{2}\text{half-life}}$	(of radioactive isotope)
UV	ultraviolet
V	volume; also electrical potential difference (p.d.); also volt — unit of p.d.
Z	atomic number

APPENDIX B

Some common hazard signs and their meaning.

 EXPLOSIVE — This substance may explode if ignited, heated, or exposed to friction or a sudden shock.

 OXIDIZING — This substance can cause fire when in contact with combustible material.

 HIGHLY FLAMMABLE — This substance may easily catch fire under normal laboratory conditions.

 CORROSIVE — This substance can destroy living tissue.

 IRRITANT — This substance causes irritation to living tissue, e.g. skin may become red or blistered after repeated contact.

 TOXIC — This substance is a serious health risk. Toxic effects may result from swallowing, inhalation or skin absorption.

 HARMFUL — This substance is less of a health risk than a TOXIC one but should still be handled with care. It may cause harm by swallowing, inhalation or skin absorption.

 RADIOACTIVE — This substance emits radioactivity and should be treated with extreme care.